The Science of MUSIC and The Music of SCIENCE

How Music Reveals Our Brain, Our Humanity, and the Cosmos

Michael J. Montague, Ph.D.

Cosmic Music, LLC

The Science of Music and the Music of Science
How Music Reveals Our Brain, Our Humanity, and the Cosmos
Michael J. Montague
Cosmic Music, LLC

Published by Cosmic Music, LLC, St. Louis, MO

Front cover artwork provided by Glenn Miller, Clayton Photos , www.ClaytonPhotos.com

Cover and Interior design: Davis Creative, www.DavisCreative.com

Library of Congress Cataloging-in-Publication Data

Library of Congress Control Number: 2019904107

Michael J. Montague

The Science of Music and the Music of Science: How Music Reveals Our Brain, Our Humanity, and the Cosmos

ISBN: 978-1-7339169-0-5 (paperback)
978-1-7339169-1-2 (case laminate)
978-1-7339169-2-9 (ebook)

Library of Congress subject headings:

1. MUS022000 MUSIC / Instruction & Study / General 2. SCI000000 SCIENCE / General
3. EDU057000 EDUCATION / Arts in Education

2019

Dedication

To E. Glenn Miller, my friend,
confidante, partner, and the talented photographer
who created the cover art for this book.

Table of Contents

Introduction

My family was not particularly musical, nor was science particularly well integrated into our family discussions. My Irish mother had a beautiful, natural, lyrical voice and she very much enjoyed singing in the church choir. My father was a funeral director, with some scientific background, but he was certainly not a scientist. Yet, thanks to their constant encouragement, I was exposed very early to music, to music making, and to science.

As a result, I've studied both science and music for more than 60 years, working as a professional scientist and as a union card-carrying musician. I've had the enormous privilege during the daytime of membership on teams of scientists who have achieved notable progress in plant science and drug discovery, favorably affecting the lives of billions of people. And then, often in the evenings, I've performed many genres of music, and composed and arranged music as well.

Throughout that time, I've observed that science and music are intimately related. Science and music—scientists and musicians—are inseparable and symbiotic. For more than 2,500 years, music has inspired scientific investigation and progress and in return science has provided musicians with untold numbers of valuable insights into their art and craft, as well as abundant, powerful technologies.

I also noticed that friends who were scientist-musicians thought about music in quite different terms from musicians who had no scientific background. Science seemed to add richness to musical understanding, though it didn't seem to increase the level of innate talent at all.

Finally, in the last few decades, science has made tremendous advances in two areas that pertain directly to our understanding of music. The first of these advances is in neuroscience, especially in the imaging of our living human brains as we perform specific tasks, and the second is in physics, especially regarding our ongoing search for the fundamental nature of reality. I was struck by how much these two advances were showing us about our humanity and about our musicality.

As an experiment, I decided to teach a short course in music and science at Washington University in St. Louis to a group of older professional people, who had come from many different walks of life. To my delight, they found the course material inspiring and useful.

They urged me to write a book. I owe my students a debt of gratitude for their participation and for their useful feedback to me.

In the light of all of this background introduction, you may well ask: who might be interested in this book? I think that if you are a parent or grandparent or anyone else who fosters the well-being of our children, this book is for you. If you are a person curious about music and how it works, not only artistically but also scientifically, this book is for you. If you are a musician, amateur or professional, interested in science or a science "geek" interested in music, this is for you also. Finally, if you are simply curious about the fundamental nature of reality and its relationship to music, as presented accessibly, this book is decidedly for you.

While perhaps hundreds of books about every imaginable aspect of music are still in print, many of them are highly academic, densely packed with facts, and not very accessible or even very understandable absent extensive background. Some, written by musicians, attempt to explain music to those who love music, but the authors give little attention to the underlying physical principles of sound and sound perception. Most of these authors don't even mention that music exists only in our brains. Some, written by scientists, are dense, mathematical, and accurate of course, but inaccessible and probably incomprehensible to non-specialists. Some, limited to a particular area, such as music and the brain, or music therapy, or the physics and mathematics of music and acoustics, fail to provide other important background information and certainly fail to provide a broad perspective.

This book is different. It aims to be an inclusive overview, one that is both accessible and easily understandable, but always accurate, without hype, exaggeration or over-simplification. After teaching the course at Washington University, I concluded that many of us are interested in how we obtain so much pleasure from music and how we can help others to experience that same pleasure. Many of us are genuinely curious about the interface between music and science. Many of us know just enough about music to be tantalized, but not enough to understand it. Many of us want to know how we can best provide musical experiences to our children or grandchildren because we've read that music helps young people to grow intellectually.

The basic architecture of this book is that we will inquire together about music and science as if we were climbing a spiral staircase. As we move up the spiral staircase, we will often return again to similar questions or ideas but at that point, we will have reached a higher level on the spiral, with more background, an entirely new perspective, and greater understanding—giving us a greater ability to see further and to grasp the concept more fully.

We'll start together with some basic definitions. What is science? What is music? Without clear working definitions, we can't gain insight. Science, in particular, is often defined inaccurately and music is often defined vaguely. We'll be clear.

We'll then proceed to the question of how music works. Our discussion won't be a thorough course in music theory, but rather a broad overview of the fundamental elements of music, and how musicians and scientists describe these elements. The descriptions made by the two groups sometimes seem to differ quite dramatically but the explanations are actually more similar than a casual inspection indicates because they deal with the same underlying phenomena. Together, we'll discover this core understanding.

We'll also discuss the instruments of the orchestra, how they're made, and how they rely upon scientific concepts for their design and function. We'll observe that without materials derived from plants, there would be no modern symphony orchestra, as we have it today with all of its power and sophistication.

Then, we'll go forward to investigate many aspects of music and contemporary biology, including:

- the neuroscience of music perception and processing,
- our musical brains,
- human musical psychology,
- how music actually changes our feelings and our physical brains,
- the clinical applications of music in promoting our well-being,
- the tools and methods used by composers to write music specifically designed to affect us in particular emotional ways,
- the evolution of musicality in our species and in other animals,
- and ultimately, perhaps most important, the critical assessment of the role of early music education in improving the intellectual capabilities of our children.

While the discussion of each of these topics is based upon the best scientific literature available, this book is written for the curious non-scientist. Good science is here, but expressed accessibly.

The final area of inquiry will be in physics, cosmology, and the fundamental nature of reality. Together, we'll begin to understand how the equations of music theory—our mathematical descriptions of music—are remarkably similar to the equations that describe our current best understanding of the fundamental nature of reality itself. We'll conduct this inquiry with minimal mathematics, but rather more historically and intuitively.

Please join me in making inquiry together. I hope that the journey will be exciting for you, providing you with entertainment, pleasure, and new knowledge.

Michael J. Montague, Ph.D.
St. Louis, Missouri, USA
29 August 2018

CHAPTER 1

What is Science?

"Imagination is more important than knowledge."
Albert Einstein
Physicist, Nobel Laureate, Violinist

Perhaps your own personal view of science is very much like the experience of many other people. You may regard science as a collection of sometimes astonishing facts about Nature. In school, you may have memorized the names of plants and animals, or the names and locations of the planets (including Pluto), or the names of the chemical elements and their properties, and hundreds of other facts. For you, that was science: a set of isolated facts, most of which you promptly forgot after the exam.

Or you may regard science as the so-called scientific method. In school, you may have been fortunate enough to do experiments or to see demonstrations of experiments and you were taught that scientists employ a specific method in doing their work. This scientific method that you learned about seemed rigid and invariant with very specific rules and procedures.

Alternatively, after thinking about it, you may have decided that science is primarily a means to an end. Science is what provides the information that we need to cure diseases, to create computers, to maintain a supply of electrical power, to communicate using our cell phones, and to enjoy many thousands of other advances and conveniences that we all take for granted in our civilization. In other words, you observed that science provides technology (know how), and you decided that technology is the *raison d'être* of science.

In point of fact, science is all of those things, but it is also much, much more, and something far more important.

Science is the single most effective, accurate, successful, reliable, and the only predictive means of inquiry that human beings have ever devised.[1]

When we scientists do our work, we nearly always begin with imaginative questions about Nature, asking ourselves these questions simply because we are curious about the Universe, and not necessarily because we think that the answers may be useful in some practical way. The questions can be about a phenomenon that is grand and huge, such as the origin of the Universe (Multiverse), or about something narrower, such as how a particular virus (*e. g.* the virus that causes AIDS) can avoid the immune system of the human body.

In many cases, we use mathematics as part of this initial inquiry, especially in physics. Mathematics is literally the language that we use to describe Nature, its laws, and its magnificence. In many respects, it is Nature's own language. Often, mathematical analysis or mathematical models lead us to insights into Nature and help us with our thinking in ways that our verbal or visual imagination is unable to provide.

In any case, we formulate our questions by using our imagination, definitely as much imagination as that required by artists or writers or composers. Our imagination must be free, unfettered, without any reliance upon authority. In science, we often have experts in a particular scientific discipline, but unlike religion, for example, we have no authorities or "authoritative" source books or "revelation." In science, no question is off limits. No direction for inquiry is impermissible or heretical.

This freedom of unfettered inquiry is part of the foundation for the success of science. If any question were off limits, or any form of authority acted as the final arbiter, or the opinions of even the most respected and knowledgeable experts weren't questioned, science would quite simply not exist. Science questions everything.

So, if science begins with open, imaginative, unfettered, authority-free inquiry, what's next?

Ideas.

Scientists get ideas about Nature as we ask our questions. Now, in the formal sense, we sometimes call these ideas hypotheses. But, frankly, that's really a formal term. More simply expressed, we get notions about how Nature might work. And, most often, these ideas turn out to be abysmally wrong—even ridiculous. We scientists must develop courage and genuine humility because we so often devise terribly wrong ideas about Nature, at least initially.

That's another characteristic of science. More often than not, we take wrong turns in our inquiry. Scientists may work on an idea for many years, only to discover that the original notion was simply incorrect. Nature can be difficult to understand!

So how do we decide whether our ideas are correct or not, that is, whether they are an accurate representation of how Nature works? We look at and study Nature, making observations to obtain:

Verifiable Evidence.

Reproducible, verifiable evidence is at the core of science. The standards for scientific evidence are very high. For example, mere eyewitness account is not acceptable, even though it's often considered a perfectly good form of evidence in a court of law. We scientists know that humans are likely to be fooled by our senses, by our expectations, by our hopes, by our preconceived notions, by our ambition, by our emotions in general. For that reason, scientists are constantly skeptical about evidence, and we always consider it to be provisional, subject to finding better evidence or to developing a different interpretation of the same evidence. In fact, good scientists know that they must constantly try to show that even their own best ideas are incorrect because, if they don't, their colleagues most certainly will.

A reliance upon obtaining verifiable evidence is called empiricism and our attitude that such evidence is always provisional is called skepticism. Empiricism and skepticism are two hallmarks of science.

So, what's the ultimate goal?

Explanation.

Our best scientific explanations all share certain characteristics. First, they are **broad**, that is, they explain a lot of observations. Second, they are **robust**, that is, they hold up to intense scrutiny. Third, they are **coherent**, that is, they seem to make good sense and fit together with our other good explanations about Nature. Fourth, they are **powerful** because they are compelling intellectually, even beautiful, based upon everything else that we understand. Fifth, they are **falsifiable**. If an explanation is not falsifiable, it simply is not a good explanation—at least not yet. Scientists spend a lot of time trying to test and to falsify explanations, because if they do falsify a very well accepted explanation, they could be awarded a Nobel Prize. [That's a little joke, of course, but winning a Nobel Prize can be a strong incentive.] Finally, and this characteristic distinguishes scientific explanations from those obtained in any other way, a good explanation is **predictive** of future events.

This ability of scientific explanations to predict future events reliably, or to predict the outcome of experiments not yet performed, even those not yet conceived, or to predict a productive new path for future inquiry (leading to important new questions) is critical. Our best formal scientific explanations are called **theories**, though recently this term has also been used less formally or rigorously to describe explanations that are not yet complete. Some of our best theories include Darwin's theory of evolution by natural selection and Einstein's theory of general relativity[2]. Both are highly predictive. Both can be tested and falsified, and every experiment, observation, or prediction about Nature made so far is consistent with and affirms those theories.

Please note that science provides only explanations, **not proofs**[3]. A proof is in the purview of logic and mathematics, where logicians and mathematicians start with axiomatic (given) statements and then use formal rules to derive a proof of a new statement. Of course, some scientific explanations are so well supported by so much verifiable evidence that they are not only highly plausible, but we think of them as true. Still, in the back of our minds, we scientists know that even the best explanation can conceivably be overturned by new or better evidence or by a new interpretation of old evidence.

Finally, with regard to this book and to our discussion of science and music, I want to emphasize that often we'll discuss findings that are very well supported by abundant verifiable evidence and good, predictive theories. For example, we'll discuss the physics of sound and the biology of the ear. Both are very well understood, based upon much evidence. We'll discuss what we think we know about the way the brain processes music but this is less understood at our point in history. We'll also discuss music and human evolution. Whereas we have much verifiable evidence about human evolution *per se*, we have only a little that specifically applies to music. Finally, we'll examine music and the fundamental nature of reality. There, the mathematics is beautiful, and much of the evidence is highly verifiable, but sometimes the interpretations of that evidence cannot yet be tested.

Science, as applied to music and to other aspects of Nature, gives us a range of levels of plausibility as a function of the quality of the verifiable evidence available to us and the status of the explanations that we possess. When our explanations are so powerful that we consider them to be true for all practical purposes, we can confidently base very important decisions upon those explanations. When our explanations are not as well-supported, however, we must acknowledge it, and become more cautious in our conclusions and decisions.

As we unravel the science of music in this book, I'll endeavor to choose the best empirical results available, constantly assessing the quality of our evidence and thereby, the reliability of our explanations.

Now, let's define the other subject of this book: Music.

CHAPTER 2

What is Music?

"If I were not a physicist, I would probably be a musician. I often think in music. I live my daydreams in music. I see my life in terms of music."

Albert Einstein

Physicist, Nobel Laureate, Violinist

One of my most cherished memories of my late father was an occasion when I was perhaps six years old. I was enamored with a particular popular song, for reasons that I didn't understand then, and don't fully understand even now when I'm 70, so many years later. All I knew for sure as a child was that the music gave me great pleasure when I listened to it on the radio. My dad learned about my infatuation, went to the record store, and bought me a 78-rpm phonograph disc of the tune. I'd bet now that he didn't understand then how much I loved him for this simple gift—this gift of music.

Just as it was for me as a six-year-old, music seems magical to most of us human beings, with its amazing, inexplicable effects on our minds. It can give us enormous pleasure, sometimes even lead us to ecstasy, change our mood instantly, trigger long forgotten memories, make us dance, often help us to connect intimately with others, and thereby allow us to share our deepest joys and sorrows.

But what is it? How can we describe music? How can we define it, even provisionally, so that we have some kind of working definition?

Music has been defined in hundreds of different ways, sometimes scientifically (physically), sometimes mathematically, sometimes philosophically, sometimes aesthetically, sometimes poetically, sometimes functionally, sometimes narrowly or very broadly. Sometimes it's even been defined quite trivially:

"Music is that which we regard as music." Martin Walker (author)

While that definition is undoubtedly accurate, it's not very useful for a discussion about music and science, is it? Can we find a more useful definition, at least as a start for our discussion together?

Perhaps a good place for us to look for a definition would be in the words of important composers, who, after all, spend their lives, their whole essence, in music. What have they said?

Beethoven

Sometimes composers can be very philosophical or metaphysical. For example, Ludwig van Beethoven said:

"Music is the one incorporeal entrance into the higher world of knowledge which comprehends mankind but which mankind cannot comprehend."

That's not very encouraging to us for this discussion because our aim here is to comprehend music at a fairly sophisticated level of understanding. Let's hope that Beethoven was incomplete in this particular opinion—or at least that we can arrive at a definition that's more concrete to meet our own needs here.

In fact, composer Igor Stravinsky was much more concrete, almost pragmatic:

"All music is nothing more than a succession of impulses that converge towards a definite point of repose."

Stravinsky

For Stravinsky, music has an aim, a "point of repose." We'll spend much effort later in this book discussing the ways that music works to create tension and then ultimately to resolve that tension, giving us "repose," especially in Chapters 5, 10 and 12.

Still, even Stravinsky's definition is not quite what we want here. Does the dictionary help at all?

The *Merriam-Webster Dictionary* defines music as:

a: the science or art of ordering tones or sounds in succession, in combination, and in temporal relationships to produce a composition having unity and continuity.

b: vocal, instrumental, or mechanical sounds having rhythm, melody, or harmony.

This definition is getting us closer. It mentions some interesting terms, *viz.* science, sounds, temporal relationship, rhythm, melody, and harmony, all of which we'll discuss at length later in this book. Still, as we'll see, phenomena exist that we would undoubtedly label as music but have no harmony or even recognizable melody.

Ideally, a working definition is as simple, pithy, and concise as we can possibly make it. How about the following?

Music is sound organized against time.[1]

That's a fairly concise definition, having only six words. But will it really work for us as we begin our discussion? [Personally, I certainly hope so, because I can't come up with a better, simpler, more succinct, or more encompassing definition.]

The first thing to check is whether anything else might fit the definition. One immediate possibility is poetry. After all, poetry relies upon sounds that are organized, and are spoken usually rhythmically over time. Our definition still seems to work, however, because poetry relies on **words**, whereas words are completely unnecessary for music. Sound alone is all that's required.

Equally important, our working definition encompasses all kinds of music, from all cultures, and of whatever style or convention. This definition makes no judgment about the quality of the music in any way and does not demand that the music be pleasurable or beautiful, although both of these highly subjective characteristics are common for music.

In thinking about this definition, three concepts immediately appear to be essential:

Sound. In this book, we'll discuss sound at length from both the scientific and the artistic perspectives.

Organized. We'll discuss the multitude of methods that musicians use to organize sound, how these methods relate to science, and how we've used science to begin to explain the methods and effects of this organization.

Time. Time is by far the most difficult and problematic concept in our definition of music. Let me explain.

While we apparently live in time and rely constantly upon its measurement, most of us couldn't begin to explain what time is or to show conclusively that it even exists as we describe it. Physicists have spent untold amounts of effort, writing untold numbers of equations, trying to understand and to explain time. While they've made considerable progress, especially during the 20th century with the work of Albert Einstein and others, mysteries still abound.

Of course, we ordinary humans are fascinated by time and make many assumptions about it based upon our common experience of past, present, and future. We can assess time intervals by using various devices that generate, measure, and display regularly recurring events. Over the centuries in civilized cultures, we've become progressively more skilled in designing and building such devices.

Probably the first observed regularly recurring external events were astronomical in origin. The sundial is an example of a clever device that measures the apparent daily movement of the sun across the sky, the result of the regular rotation of the earth. Hourglasses measure the regular recurrent falling of sand grains through a small orifice under the influence of the earth's gravity and can be constructed to coincide with astronomical units of time.

Galilei

The great Italian Renaissance scientist, Galileo Galilei, whose father, by the way, was a musician, studied the movement of pendulums and discovered that they swung regularly, recurrently, and predictably. Apparently, Galileo's first observation in this regard was a swinging cathedral chandelier, which he found to be synchronous with his own heartbeat. Of course, the regular movement of a pendulum was exactly what was required for a clock, and Galileo's observations inspired its invention. Today, our most accurate clocks measure other kinds of regularly recurring events, such as the vibration of an electrified quartz crystal (our wrist watches)[2] or rely upon the natural vibration frequency of the energy emission of a particular element, most often cesium or ytterbium (atomic clock).

As we shall see in the next chapter of this book, music relies upon the human ability to measure time intervals internally in our own brains, though rarely as accurately as a good clock. Musicians often refer to the utilization of this innate ability as **keeping time** and some people are much better at this task than others. Sometimes we musicians get in the bad habit of assisting our brains with keeping time by tapping our foot "in time," though this practice can be very distracting to the audience. Foot tapping, however, does reflect the strong connection between music and movement. Professional musicians know that if you must tap to keep time, tap your toe inside your shoe! We'll discuss this ability to keep time and its importance to music in a fair amount of detail in Chapter 3, as well as later in this book when we discuss the processing of music in our brains (Chapter 9).

But before we go on, I would be remiss if I left you with the impression that time is all about regularly recurrent events and their measurement. I'd be even more remiss if I left you with the impression that physicists think of time only in that way, or even agree that the flow of time actually occurs, rather than being merely illusory—or certainly if I left you thinking that time flows at the same rate everywhere in the Universe.

Philosophers and scientists have been debating the definition and nature of time for thousands of years. The great 17th century physicist and mathematician, Sir Isaac Newton, thought that time flowed at the same rate regardless of one's location in space and even in the absence of a succession of physical events. In other words, time simply existed as a constant, immutable, invariant, fundamental property of the Cosmos. Newton's ideas about time are completely intuitive, in accord with our common, everyday experience.

Einstein

Einstein[3] changed all of that in the early 20th century with several critical insights. First, he showed that time is **relative.** Intervals of time differ depending upon location and speed of movement. The passage of time slows down as an object moves more rapidly, and would stop flowing completely in an object if it could reach the speed of light. All of this comes directly from the observation that the speed of light is constant and finite.

Einstein also showed that space and time constitute one **physical** thing, fused together, which he referred to as space-time with four dimensions. He developed an entirely new theory of gravity that differed dramatically from Newton's conclusions. Einstein showed us that our perception of gravity results from the presence of mass (matter) as it actually distorts space-time. He also demonstrated a profound relationship between motion through space and the passage of time. This dilation of time (the slowing of time with increasing speed and with the gravitational curvature of space-time) requires that the clocks in the GPS (global positioning system) satellites orbiting above us must be corrected for these effects. Without such correction, your car's GPS would quickly become worthless as a location or navigation device. After only 24 hours, the GPS measurement would be off target by seven miles.

Finally, the most counter-intuitive idea of all is that our commonly held concept of time flowing from past to present to future is actually merely an illusion, however persistent this illusion may be in our minds and in our lives. Einstein's mathematics shows us that the past, the present, and the future simply exist simultaneously—always, eternally, as if frozen.

We'll discuss these insights from another perspective in the last chapter of this book as we learn more about what physicists and cosmologists think about the fundamental nature of reality and its rather surprising relationship to music. No one really understands time. Our best idea is merely descriptive: time is the quality of reality that allows for change.

But to close this chapter with some comic relief to ease the pain of destroying your illusions about time, let me quote Hector Berlioz, another famous composer:

"Time is a great teacher, but unfortunately it kills all its pupils."

CHAPTER 3

The Pulse

"Think you I bear the shears of destiny?
Have I commandment on the pulse of life?"
William Shakespeare
***King Lear*, Act IV**

Ah, the pulse of life. We are all familiar with the sensation so easily felt in our wrist or in our neck beneath our jaw—the regular, recurrent expansion that we all know does indeed sustain our very life.

Nature is filled with many regular pulsations and movements. They are found everywhere. Scientists have concluded that such regular recurrence, or periodicity, is a theme common throughout all of Nature.

Perhaps one of the most interesting of these natural phenomena is the pulsar, a neutron star that rotates very rapidly and very regularly. As it spins, a burst of electromagnetic radiation is emitted directly toward the earth with a precise periodicity. Astronomers have used pulsars to document the existence of planets existing outside of our solar system and some pulsars rival the time-keeping accuracy of atomic clocks.

The pulse is also fundamental to being human. Besides our heartbeat, think of all of the other physiological functions of our bodies that are periodic. Perhaps you immediately think of breathing or the sleep-wake cycle or the menstrual cycle, which tends to synchronize automatically when women live together in close proximity. Less apparent to most of us are the regular pulsations of our intestines, as they move food from our stomach through the alimentary tract, or the circadian patterns—from the Latin *circa* (approximately) and *dies* (day)—of many of the biochemical reactions and changing hormone secretions at the core of our biology, most of which must be reset when we move through widely divergent time zones. Our bodies automatically use the light-dark cues of our environment to entrain them, even though that entrainment doesn't seem to occur nearly rapidly enough for many of us who travel frequently.[1]

The pulse is fundamental to music as well. Musicians must "feel" the pulse of the music, as must the audience. When rehearsing a musical passage, it is not uncommon for conductors to point out to the orchestra the location of the musical pulse, whether in the bassoon or in the celli or often in the basses and tympani. By that statement, the conductor is indicating to everyone in the orchestra that each musician must be aware of the pulse so that the musical performance is coherent, so that it "hangs together," and so that each individual part of the music aligns with every other part. The pulse is fundamental to the measurement of time by each musician and serves as an organizing principle for the sound that each makes.

Some theorists would argue that certain forms of music, such as Gregorian chant, an ancient form still used in liturgical services today, has no pulse.[2] They would say that it's free form. While I'd encourage your own assessment (there are many recordings available online), I think that you'd find that the pulse remains present in Gregorian chant and in other forms of plainchant. Often the pulse is merely implied, arising from the sung words themselves, with subtle lengthening or shortening of a syllable. And the pulse certainly can vary in regularity more than in many other forms of music. The bottom line, though, is that the pulse remains as an organizing principle.

In any case, all of this points to an amazing faculty of humans (and apparently a few other animals, which we'll discuss later in Chapter 13), namely, the ability to entrain to a pulse. Most of us have enjoyed the pleasure of clapping our hands in unison with a group of other people or swaying together as a crowd with a constant periodicity that is set by one or more members of the group. It's a delightful feeling of being bonded to our fellow humans, connected so cohesively.

Humans have the ability to entrain to a pulse over a wide range of periodicities from about 40 pulses per minute to perhaps 180, mostly limited by muscular coordination, and perhaps the neuronal circuits in our brains that are devoted to such entrainment (more about that later as well in Chapters 9, 10, and 11). Our ability is very plastic in this regard and we can easily entrain to a new period, either one set externally by another person, or one set internally ourselves.

The Pulse and Rhythm

While the pulse *per se* is critical to music and to music making, musicians also use a variety of specific terms to communicate aspects of the pulse to each other. It's important to remember that the pulse has no intrinsic characteristic other than periodicity. Music is much more than mere periodicity, of course, and musicians use the following terms to add critical distinctions:

The Beat: Musicians use the term "beat" to group the recurrent pulse and to give it meaning. While there's no distinction between one abstract pulse and the next, beats vary significantly. Some are heavier, that is, they receive more emphasis. Some are lighter, and they receive less emphasis. Beats are grouped together, most often in 2's, 3's, 4's, 6's, 8's, and 12's, though sometimes in 5's, 7's, 9's, or even 11's and 13's (rarely). Groups of beats are placed into segments of time called measures, and each measure is separated from the next by a vertical bar line in Western musical notation. While the measure (the unit of time) is distinct from the bar (the line that indicates the beginning and end of a measure), musicians will often use the terms interchangeably to describe the basic unit of time employed in a piece of music. If the music is written down, the bars are often numbered to aid the rehearsal process.

The Tempo: Taken from the Italian, meaning time (plural: tempi), musicians use this term to describe the speed of the music, that is, how rapidly one beat follows another. It is usually indicated as beats per minute (bpm). Contemporary composers will often designate a tempo by writing a metronome marking in beats per minute, specifying the note symbol that receives one beat. A metronome is basically a mechanical or electronic clock, a time keeper, which usually clicks (loudly) with each pulse. Before the advent of accurate clocks, musicians had no choice but to use various written words to convey the tempi desired. Because many important early composers were part of the Italian Renaissance and used Italian terms, these terms are still often in Italian, and include:

Lento: Very slowly

Larghetto: Slightly faster than *Lento*

Adagio: Slowly

Andante: A walking tempo

Moderato: Speed walking

Allegro: Fast

Presto: Very fast

Prestissimo: Inhumanely fast, as fast as the musician can play it

Obviously, such descriptive terms are much less exact than specifying the number of beats per minute, and the interpretation of such tempo markings often creates consternation and intense disagreement among musicians, in particular among conductors who insist that their interpretation, and only their interpretation, could possibly be the one desired by the long-dead composer. In effect, one could say that time is relative for musicians, too.

Rhythm: The definition of rhythm is layered. The most basic definition is the pattern of emphasized (heavy / strong) and non-emphasized (light / weak) beats. For example, in a standard waltz, the rhythm is 3 beats per measure, with a pattern of **HEAVY**-light-light / **HEAVY**-light-light, etc. Expressed another way: **ONE**-two-three / **ONE**-two-three, etc. But rhythm can be far more complex than that because it also describes the pattern and length of sounds, the subdivision of the beat, and the pattern of silences as well as sounds. One other important idea is the use of **syncopation**, which occurs when the emphasis is placed on the off-beat, the beat that would normally **not** receive emphasis. Syncopation is a critical technique in jazz and much popular music, giving it great rhythmic interest and momentum. One thrilling rhythmic technique, found particularly in African music, is called **cross rhythm**. For example, a 2-beat pattern may be superimposed over a 3-beat pattern, providing a wonderful sense of tension and excitement.

Meter: Meter describes the recurring pattern of rhythm in the music. In the waltz example above, we have a triple meter. Marches are in duple meter (**LEFT**-right / **LEFT**-right, etc.). Much popular music is written in quadruple meter. In fact, quadruple meter is so common that it is often abbreviated with a "C" meaning common time. In quadruple meter, the pattern of emphasis is often that the first beat of the measure receives the greatest emphasis, the second much less, the third receives more emphasis than the second, but not as much as the first, and the fourth beat of the measure receives the least. We could represent this pattern as: **ONE** – two – **three** – four / **ONE** – two – **three** – four. A syncopated rhythm in common time, then, could be One – **Two** – three – **FOUR.** If you've ever observed a crowd clapping in rhythm to a jazzy tune, you've undoubted seen and heard them clapping on the off-beat.

Time Signature: The time signature allows the composer to communicate the meter to the performer. It is written as two numbers, one above the other. The top number describes the number of beats per measure and the bottom number indicates which note symbol is assigned one beat. In common time, indicated as 4 / 4 time, the top number shows the performer that there are four beats per measure, and the bottom number that the quarter note is assigned one beat.

Codification of these definitions and accepted practices, together with reasonably accurate musical notation, is a quite recent event in human history. These concepts are by no means universal and non-Western cultures often make perfectly good music absent any notation or alternatively, employ an entirely different form of notation. We will discuss non-Western musical systems later in the book (primarily in Chapter 13). Our initial emphasis on Western traditions is not meant in any way as a critique of other systems. Western musical conventions are a good place to start, however, if only because these conventions have penetrated throughout much of the globe.

One final bit about the notation of rhythm involves the symbols that Western musicians use to designate how the beat is divided in time and which portion of that division is used to make a sound or to remain silent. The main concept here is that the beat ostensibly can be divided into a virtually infinite number of sectors, though a practical limit is reached quickly because of the limitations of our human brains and muscles. That practical limit, generally, is to divide the beat into no more than 32 sectors, even at very slow tempi. All of this is akin to the way scientists divide a second of time into tenths, hundredths, thousandths, millionths, billionths, and even trillionths of a second and smaller—but scientists use electronic devices to make these extremely fine distinctions possible, not their senses.

Let's use common time, 4/4 time, to illustrate. Here are the symbols that we use to indicate the length (fraction of the beat) designated for sound production, or for silence (the rest marking at the far right of the chart.)

Symbol	Name	Number per measure (4/4)	Rest
𝅝	whole note	𝅝 1 per measure	𝄻
𝅗𝅥	half note	𝅗𝅥 𝅗𝅥 2 per measure	𝄼
♩	quarter note	♩ ♩ ♩ ♩ 4 per measure	𝄽
♪	eighth note	♫♫ ♫♫ 8 per measure	𝄾
𝅘𝅥𝅯	sixteenth note	𝅘𝅥𝅯𝅘𝅥𝅯𝅘𝅥𝅯𝅘𝅥𝅯 𝅘𝅥𝅯𝅘𝅥𝅯𝅘𝅥𝅯𝅘𝅥𝅯 𝅘𝅥𝅯𝅘𝅥𝅯𝅘𝅥𝅯𝅘𝅥𝅯 𝅘𝅥𝅯𝅘𝅥𝅯𝅘𝅥𝅯𝅘𝅥𝅯 16 per measure	𝄿

By using a combination of note symbols and rest (indicating silence) symbols, the composer can specify extremely complex rhythms. Of course, it's the task of the performer to execute those directions accurately and beautifully. This chart is by no means complete because there are other notations that composers use to designate the division of the beat into multiples of 3, or 5, or 7 as well. Other symbols are used to show emphasis or length of sound. Also, please remember that no matter how good musical notation may ever be, it is always an **approximation.** Musicians rely upon their training and experience to interpret these symbols in the context of musical style and convention to approach the intentions of the composer.

In brief:

1. The pulse is the time-keeping element of music and professional musicians can maintain an accurate internal pulse or entrain their own pulse to the pulse of the musical group;
2. The periodic, recurrent, abstract pulse is used to define the beat, which is assigned a range of greater and lesser emphasis. This allows the metrical grouping of the pulse into units of time called measures, separated by vertical bar lines in Western notation;
3. The speed at which the beat occurs is called the tempo, and is most accurately specified as beats per minute;
4. The rhythmic pattern of sound and silence can be notated quite well, but always approximately, using standard conventions and symbols.

In later chapters of this book, we will discuss the ways in which deviations from this regularity contribute to the emotional evocation wrought by music and how all of this works neurologically and psychologically (Chapter 12). But for now, let's move on to the next element of music: sound.

CHAPTER 4

Sound

"All sounds of the earth are like music."
Oscar Hammerstein II
American Librettist and Lyricist

I spent my boyhood in Michigan, a state surrounded by and covered with freshwater lakes. During those all too brief Michigan summers, my dad and I went fishing for bluegill and perch fairly often, boating on deep lakes, with the sunlight glistening as it reflected from the water. When we were fishing, I had time to observe and to think. I noticed that the ripples caused by a pebble tossed into relatively still water were amazing and beautiful. Small waves moved outward in a perfect circle from the point where the pebble entered the water and then gradually faded away to nothing. The pattern of waves created by several pebbles tossed into the water simultaneously, as the ripples mingled and crisscrossed, was even more complex and a bit mysterious.

When I later studied physics formally, I found out much more about waves and discovered that they exist everywhere in Nature. In fact, all types of waves share some commonalities in their mathematical descriptions—quite noteworthy, but not necessarily unexpected. We'll discuss this in greater detail towards the end of this book.

Physicists define waves as the propagation of a disturbance. Those water waves that I saw as a boy on the lake were the propagation of the disturbance caused by the pebble as it entered the water. The pebble disturbed the molecules of water. The propagation of that energy—the predictable movement outward of the disturbance—occurred because the molecules of water that were initially disturbed by the pebble then in turn disturbed other molecules and they disturbed other molecules, etc. As the disturbance spread out, the energy of the disturbance eventually died out as it became distributed over more and more molecules. And gradually the ripples disappeared.

Physicists divide waves into two primary types. The first is called a *transverse* wave. In this case, the disturbance is propagated at a right angle to the disturbance. The water waves that I saw on the lake in Michigan are primarily transverse waves. Think of it this way: the pebble enters the water vertically, approximately at a right angle to the surface of the water. The disturbance is propagated visibly at the surface of the water, at a right angle to an imaginary line of entrance of the pebble. The water wave is characterized by a series of crests (peaks) and troughs on the surface of the water, meaning that the height of the water visibly changes.

Sound Waves

Sound, on the other hand, though certainly also the result of a disturbance, propagates as a **longitudinal wave** and only through a medium. A longitudinal wave propagates the disturbance in a direction parallel to the disturbance, rather than at a right angle to the disturbance.

Sound waves are **compression** (sometimes called pressure) waves. Let's assume that air is our medium, because music, as sound, is ordinarily propagated to our ears in the medium of air. With sound, the disturbance is initiated by something that **vibrates** that is, something that moves back and forth, thereby alternately **compressing and rarefying** the molecules of gas next to it.

Given how abstract all of this description may seem, it's definitely an opportunity for an illustration. Here is a diagram of a loud speaker. If you've ever touched the cone of a loudspeaker while it was making sound, you felt a buzzing, a tickle, in your fingers. That's the vibration designed by audio engineers to disturb the air molecules precisely and accurately.

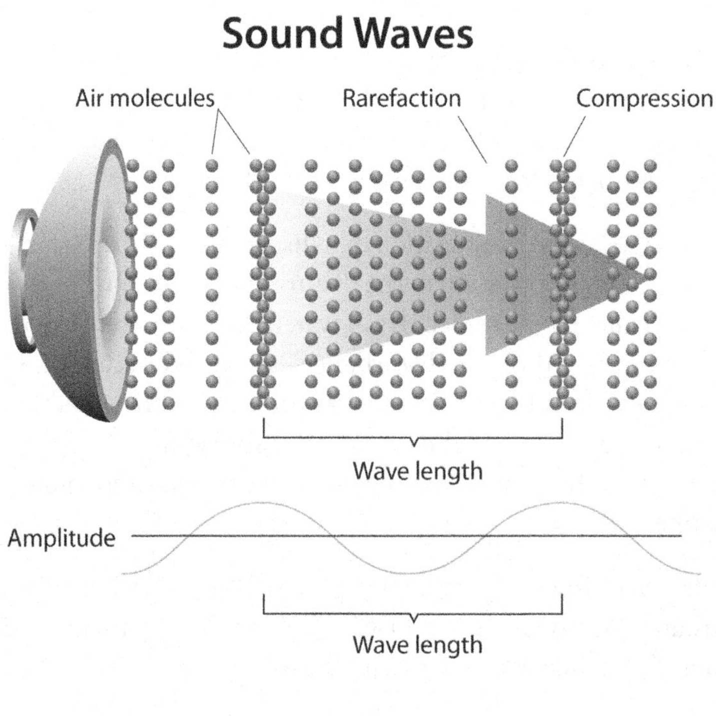

The diagram also shows two descriptive characteristics of waves. The first is the **wavelength,** the linear distance between compressions, which equals the distance from one rarefaction to the next. The second is the **amplitude** of the wave. In the case of sound, the amplitude of the wave can be thought of as the degree to which the air molecules are compressed and rarefied. With low amplitude, the molecules are barely compressed, whereas with high amplitude, the molecules are significantly more compressed.

One characteristic of waves not shown in this diagram is the **frequency** of the wave. The frequency is defined as the number of entire waves (compression to compression or rarefaction to rarefaction) that occur in a given unit of time. Mathematically, it's the number of complete cycles divided by a defined unit of time. Physicists most often express the frequency of a sound wave as cycles per second. Rather than use an abbreviation such as cps, however, scientists have chosen to name this measurement after the 19th century German physicist who first conclusively demonstrated the existence of the electromagnetic wave. His name was Heinrich Hertz, and physicists express the frequencies of many types of waves, including sound waves, in Hz, meaning cycles per second, or kHz, meaning thousands of cycles per second, or MHz, meaning millions of cycles per second, etc.[1]

Physicists and Musicians

Now, to return to music, we musicians express these characteristics of sound waves in terms that make sense to us. But the terms are different from those of physicists.

A physicist would say:	**A musician would say:**
Frequency	***Pitch (or sometimes: tone or note)***
Amplitude	***Loudness (or usually: the dynamic)***

Finally, referring again to our loudspeaker diagram, the compression-rarefaction cycle is also represented there with a graph of a sound wave that might be made by a tuning fork. This uncomplicated wave, having a simple form and periodicity[2], is sometimes called a sine wave because it's described mathematically using the trigonometric function of the same name. The mathematical representation is quite beautiful and we see a similar mathematical representation as we investigate any phenomenon of vibration. Please remember, however, that sound waves are longitudinal compression waves and do not **physically** display crests (peaks) and troughs as do water waves. With sound waves, the crests and troughs come from the mathematical description, not from the physical reality of sound.

A Few Numbers About Sound Waves

For clarity, I think that it would be useful to add some quantification to these concepts. First of all, we learned above that sound waves are propagated only through a medium. Electromagnetic waves, such as radio waves, require no medium, and they propagate much faster at approximately 300 million meters per second (186,000 miles per second). Yet, the mathematics used to describe them is very similar to that for sound waves.

In contrast, the speed of propagation of sound waves through dry air at 20°C (68° F) is 343 meters per second (767 miles per hour). Of course, the speed of sound is a function of the medium. In fact, it increases with rising air temperature, and somewhat with increasing relative humidity. Musicians know that higher temperatures and changing relative humidity in a concert hall change the sound of the orchestra for many reasons, including the speed of sound within their wind instruments.[3]

The loudspeaker diagram above is misleading in terms of its dimensions. The actual amount of compression of air molecules with ordinary sound is minute. Tiny. The human ear is capable of detecting a sound pressure wave of one-billionth of the pressure of the atmosphere as it presses on our bodies. In other words, our ears are extremely sensitive to small pressure displacements in air. Another way to express this is that even the very loud sound pressure wave generated by a jet engine causes the molecules of air to be displaced only by about 0.01 mm (0.004 inches). In the same wave, the pressure at compression is only about 0.0003 atmospheres above the atmospheric pressure of 1 atm.

Also, as a point of calibration for our thinking, let's consider the wave generated by a piano when middle C is sounded. The frequency of middle C is 261.6 Hz. This corresponds to a wavelength of 1.3 meters (51 inches). Don't let the loudspeaker diagram mislead you. Even the highest note on the piano at 4,186 Hz has a wavelength of 3.23 inches.

The Vibrating String

Lest I continue to bore you with numbers, let's move on to vibration, the source of sound. We'll use a vibrating string for our initial analysis. In that regard, we become students of Pythagoras, the sixth century B.C.E. natural philosopher and mathematician who was fascinated with music and vibrating strings. While best known for his Pythagorean Theorem (the sum of the squares of the sides of a right triangle is equal to the square of the hypotenuse), his contribution to music theory was enormous.

The work of Pythagoras is an example where intense curiosity about music and how it works led scientists and mathematicians to discover fundamental properties of Nature. It was the curiosity of Pythagoras and other philosophers and scientists interested in music that contributed to a profound understanding of waves and our ability to predict their behavior. This understanding not only helps us in practical matters such as in predicting the wave height created by a Tsunami, but it even extends to the equations for the fundamental nature of reality itself. We'll learn much more about this in the last chapter of this book.

When we pluck a single, taut string, it begins to move back and forth, that is, to vibrate, restrained by the fixed attachments at its ends. The fundamental frequency of vibration is a function of its length, its mass, and the tension on the string. There's an easy algebraic formula that describes the relationship between these three variables.[4] String instruments are designed with strings of the correct mass and length. Musicians "tune" their instruments by adjusting the tension on the string. A "tighter" string will produce a higher fundamental frequency of vibration than a "looser" string of the same mass and length.

As a practical matter in music-making, the **tuning** of strings, and in fact adjusting the tuning of all instruments that are not fixed pitch, is required so that the musicians in the orchestra play in tune, that is, they agree on a standard frequency for any given note. Musicians call this **good intonation**, and often argue among themselves about it. Musicians hear bad intonation as an ugly sound. The sound waves are out of phase and the compressions and rarefactions do not align because the musicians are producing slightly different frequencies. This results in the sensation of **beatings**—a sort of quivering—in the sound, an unpleasant sensation of alternating louder and softer sound that results from the rapid alternation between compressions / rarefactions reinforcing or canceling each other.[5]

The analysis of a vibrating string is actually a bit more complicated than the consideration of only one vibration frequency. While the string has a fundamental, natural frequency of vibration, it **simultaneously** vibrates at multiples of that frequency: 2x, 3x, 4x, 5x, 6x, etc. as illustrated on the next page:

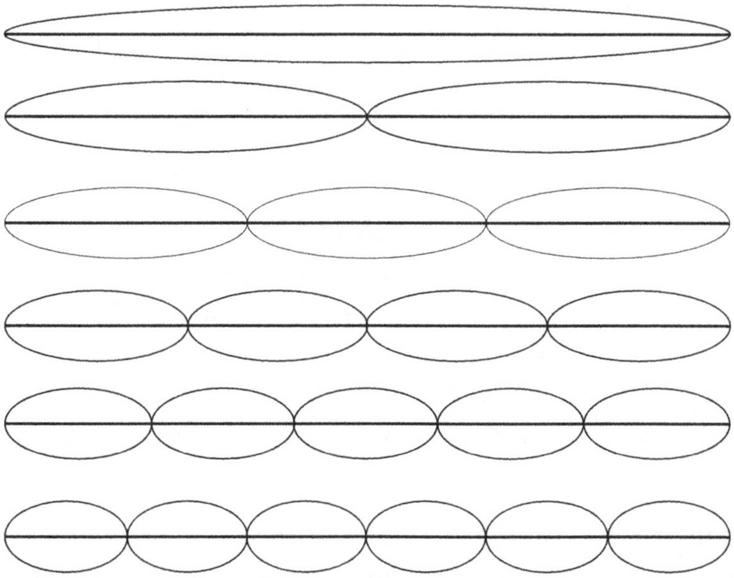

A similar phenomenon is observed when columns of air are set in vibration in the tubes of wind instruments, something that we'll discuss later in more detail. This is called the **harmonic series** and is an important concept in music theory.

Now, depending on the particular source of vibration, the **relative intensity** of each of these vibrations can vary widely. The ultimate sound that reaches our ears is a combination of the intensities of each of these harmonic vibrations. This summation produces a final waveform that is quite different from the simple sine wave illustrated with our loudspeaker. While still maintaining the frequency and wavelength of the fundamental harmonic, the sound is colored by the unique intensities of the other vibrations. Each musical instrument, and each human voice, produces a characteristic mix of these harmonics. The mix of harmonics enables us to discern the voice of one person from another.[6]

We refer to this characteristic of sound—this tone color—as **timbre** (pronounced: *TAM-ber,* or if you speak French, *TAM*-bruh). It's what makes a piano sound different from a clarinet, which sounds different from a violin.[7]

The final characteristic of musical sound that we should mention here because it distinguishes the sound of one instrument from another is related to the intensity and quality of the sound from the time it is initiated on the instrument to the time it has decayed away passively or is actively stopped by the musician. This is called the **time envelope** of the sound. For example, the string on a piano is struck with a relatively soft hammer to initiate the sound, which then has a characteristic rate of decay. A plucked harpsichord string has a very different time-related profile, though both the piano and the harpsichord are stringed, keyboard instruments.

Musical Instruments and Sound Characteristics

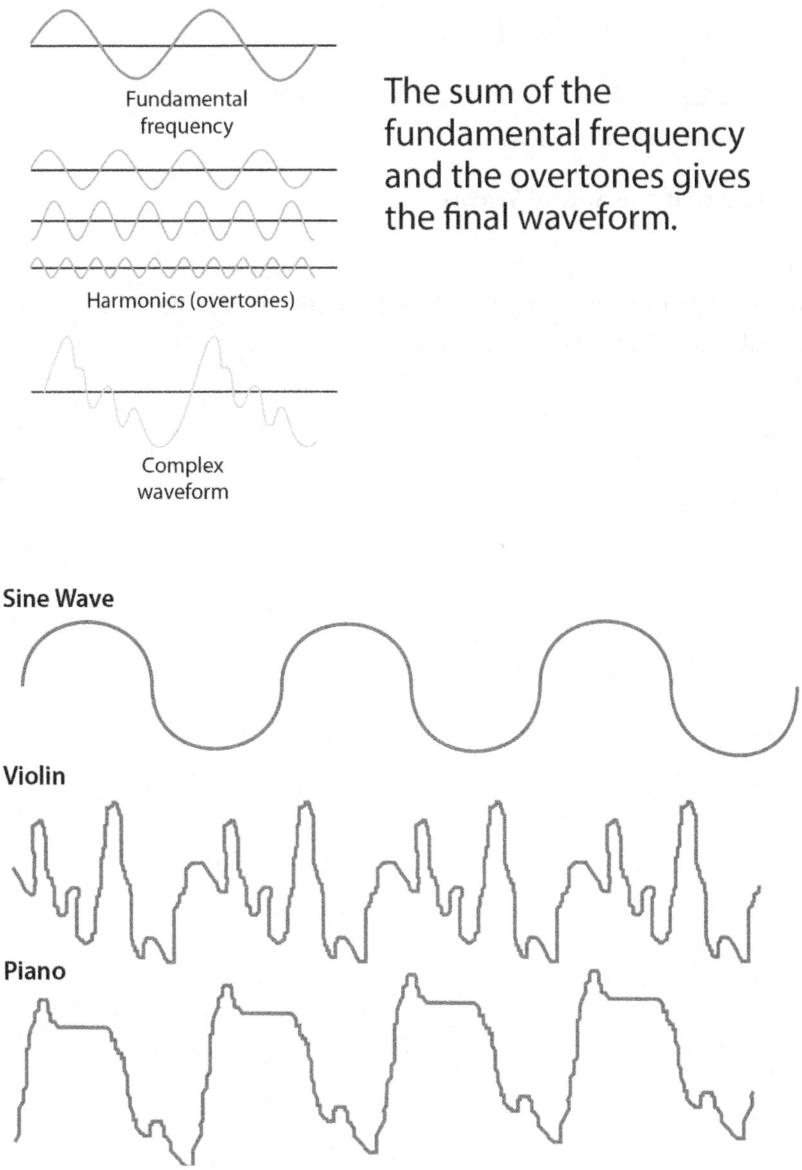

Many scholars group the time envelope quality as part of timbre, and that makes much sense for simplicity, but we should realize that one of the reasons that electronic instruments (electric piano) never quite match the sound of the acoustic instrument (real piano) is that they fail to match the time envelope accurately.

In summary, then, musicians describe musical sound as possessing the following qualities, each very well explained by our scientific theories of sound waves:

Pitch (The Frequency of the Sound Wave)

Loudness (The Amplitude of the Sound Wave). Note: loudness is most often described by musicians as the **dynamic** of the music.

Timbre (The Form of the Sound Wave)

These qualities are critical components of our musical experience. And now that we understand the components of musical sound, let's move on to asking questions about the higher-level organizational elements of music itself.

CHAPTER 5

Melody

"Melody is the essence of music..."
Wolfgang Amadeus Mozart
Composer

"A melody is not merely something you can hum."
Aaron Copland
Composer

I sincerely hope that you too enjoyed the great good fortune that I enjoyed as a very young boy: a mother who sang to you. My auburn-haired, quintessentially Irish mom possessed a lovely, light soprano, quite typical of many an Irish lass. Without any formal musical training whatever, she had what we musicians call a "good ear." She sang perfectly in tune *a cappella*, that is, without instrumental accompaniment. Some of my earliest memories are of my mom singing lullabies and popular songs to me. At that time, in the early 1950's, she could have no understanding of the scientific studies that we would do decades later showing the value of mothers singing to their babies. She just knew intuitively that this was a good thing. And she, herself, I suspect, thoroughly enjoyed it as well.

Besides her obvious love for me, what I remember most was the melody of the songs. We sometimes think of melody as **the tune**[1] and that's not an incorrect notion at all, in particular for lullabies and popular songs, where the melody is so prominent. But I think that we have to be much more precise as we think about melody, in part because it can assume so many forms.

Before we can even begin to inquire about the nature of melody, something so important to our pleasure in music, we must learn some more basic music theory. I'm convinced that greater understanding of the theory will lead to greater appreciation of the concepts and will be well worth the effort. Please bear with me. I'll try to make it as painless as possible. [Aside: I know that some readers are curious about music but haven't had exposure to many basic theoretical concepts. Some of these concepts are not intuitive and I'd like to help to make them as easy and accessible as possible. But for those who find this a review, feel free to skip to Chapter 6.]

Vibrating Strings Reprised and Some Basic Music Theory

Let's return to vibrating strings. Musicians have always been challenged to make plucked strings sound pleasant. There's nothing worse than an ugly sounding string, except if it's 2 or 3 or 4 or more ugly sounding strings.

One of the first things that ancient musical humans undoubtedly discovered was what I call the wonder of the **octave**. We'll learn some biology about this when we discuss the perception and processing of music by our ears and brains, but for now let's simply observe that the octave is a very special relationship between pitches. So, what is it?

When a frequency (pitch) of a new sound is precisely a multiple of 2 (2x, 4x, 8x, etc.) or a fraction of 2 (x/2, x/4, x/8, etc.) of the original pitch (where x is the original frequency), it sounds to the ear as the same pitch, only higher or lower, but matching it. The two pitches sound perfect together, virtually indistinguishable, perfectly in tune. Incidentally, this is approximately the pitch difference between the voices of adult human males and the voices of women and prepubescent children. This 2-fold difference in frequency, or ½—fold difference, whichever way you care to think about it, is called the octave.

How do we produce the pitch an octave higher on a vibrating string? If I were there to demonstrate for you, we'd find that I could simply reduce the length of the string to ½ of its original length, placing my finger in the middle of the string and plucking it on either side of my finger to produce the octave of the original pitch. A pitch one octave higher yet is obtained by reducing the newly shortened string to ½ its length (¼ the length of the original string). In other words, reducing the length of the string by ½ causes it to produce a frequency 2-fold higher than the original frequency.

Do you remember the harmonic series from Chapter 4? Take a look at the drawing that illustrates it. Note that the octave is the second harmonic (2x) above the fundamental frequency of the string. The octave is physically represented as a vibrational mode of the vibrating string. But, as we observed, many other sound frequencies are also there: 3x, 4x, 5x, 6x, etc. where x is the fundamental frequency of the vibrating string. Note: the harmonic series of frequencies is actually produced by integer (whole number) divisions of the string. In other words, these pitches can be produced by dividing the string into thirds (the third harmonic), into fourths (the fourth harmonic), into fifths (the fifth harmonic), into sixths (the sixth harmonic), etc. just as we did with the octave (the second harmonic).

While we can't be certain about the historic development of stringed instruments, we can reasonably speculate that ancient musicians had multi-stringed instruments that produced pitches dividing the octave into intervals. These pitches formed a **scale.**[2] Musical scales are made from consecutive frequencies that lie between two frequencies that are one octave apart. In other words, a scale is a set of pitches that **divides the octave into intervals. Important: Melodies are made of the pitches of scales.**

Types of Scales

The first such division of the octave widely employed by humans was probably the **pentatonic scale** with five (*penta*) intervals, 5 pitches + the octave. While there are several versions of the pentatonic scale, one form or another remains in use today in much of the world, particularly in Asia. The pitches of a common form of the pentatonic scale are all expressed as ratios of simple integers. If we set the initial frequency as the pitch of middle C on the piano, the frequencies would be:

(Initial Frequency of Middle C) x **(1/1)** = 261.626 Hz

(Initial Frequency of Middle C) x **(9/8)** = 294.329 Hz

(Initial Frequency of Middle C) x **(4/3)** = 348.835 Hz

(Initial Frequency of Middle C) x **(3/2)** = 392.439 Hz

(Initial Frequency of Middle C) x **(5/3)** = 436.043 Hz

(Initial Frequency of Middle C) x **(2/1)** = 523.252 Hz

One of the advantages of the pentatonic scale is that all of the intervals sound pleasant when played together. It seems to the listener as if this is a "natural" division of the octave. The pentatonic scale is occasionally used today in Western music in well-known tunes such as *Amazing Grace* and the Scots folk song, *Auld Lang Syne.*

Our old friend from Chapter 4, Greek mathematician and natural philosopher Pythagoras, was fascinated by vibrating strings, by music, and by mathematics. As he studied this pentatonic scale, he noticed that these pleasant intervals were simple whole number (integer) ratios multiplied by the initial frequency, as reflected in the length of the string (as a reciprocal, that is, 1 / frequency) because shorter strings produce higher fundamental frequencies. For Pythagoras, this whole number ratio relationship was a delightful discovery. He posited that this relationship was basic to the nature of reality and that the ratios reflected the beauty and the order of Nature itself.

To demonstrate how this works out, we can take a single string, fixed at both ends, and then place our finger on the string at various points to stop it from vibrating at that point. If we pluck the remaining portion of the string, which is now shorter than the original string, it will produce a higher fundamental pitch. This is the principle used by musicians who play violin, viola, cello, and double bass.

For the pentatonic scale:

^ = the position your finger would touch (stop) the sting, and then you'd pluck on the left.

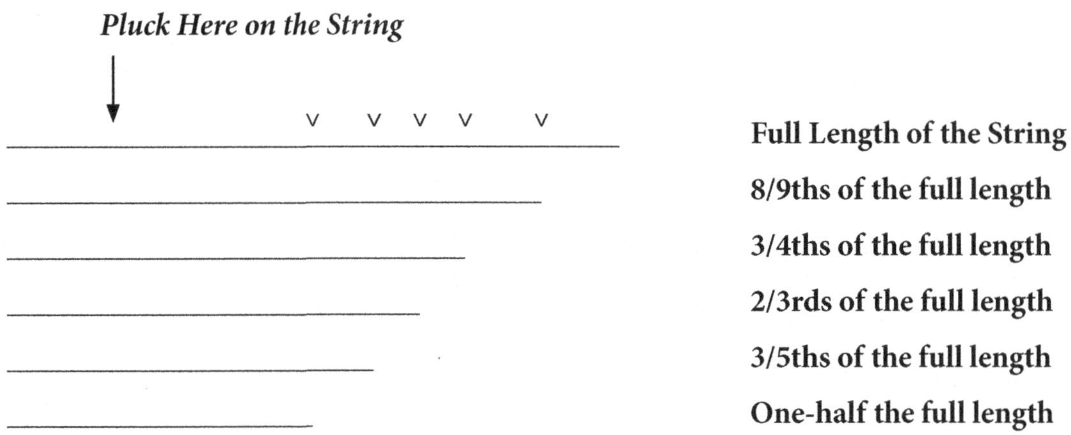

The genius of Pythagoras (and his followers) has enormously influenced Western music even to the present day with his invention of another scale based solely on the octave (2/1) and a succession of perfect fifths, which are defined as the initial frequency multiplied by 3/2. The scale is called the **diatonic scale** and is very familiar to us who are immersed in Western music. It is a scale with 7 intervals and 8 pitches including the octave, which are in turn a subset of pitches selected from a scale that divides the octave into 12 intervals and 13 pitches including the octave, called the **chromatic scale.**

Let's stop for a moment here and summarize:

"Generic" Scale	Consecutive pitches that divide the octave
Pentatonic Scale	Classically, five consecutive pitches, defined by simple whole number ratios
Diatonic Scale	Example: the "Do-Re-Mi-Fa-Sol-La-Ti-Do" scale; adds Mi and Ti to the pentatonic
Chromatic Scale	Divides the octave into 12 intervals

Main Point Reprised: The reason that we learned all of this stuff about scales is that melodies are made up from a succession of pitches taken (at least primarily) from a particular scale.

The Science of Music Notation[3]

For untold numbers of centuries, humans had no method for notating (recording in written form) the music that they made with any accuracy or in a way that enabled other musicians to perform the same music. Even the brilliant Greeks failed to devise a system of notation that continued to be used by musicians. Various cultures have devised many different methods that were basically mnemonic, based upon symbols that served as a reminder of music that the performer already knew quite well. Unless the performer had already heard the music, it is unlikely that he or she could read the symbols and reproduce the piece of music with fidelity. That meant, of course, that music was largely preserved aurally rather than symbolically.

If we think about music notation from a scientific perspective, however, we could well imagine that it might be graphically represented with something like a Cartesian graph. René Descartes, the famous 17th century French mathematician, devised this system for the graphical representation of information well after music notation had been invented in Western Europe, but it serves to help us to understand some basic principles.

A Cartesian graph starts with two axes, often labeled simply x and y.

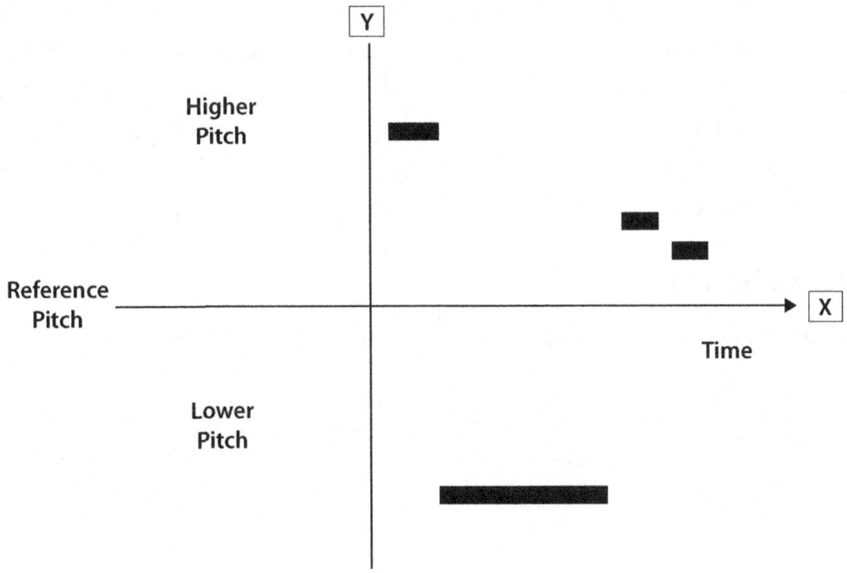

In this case, the y axis represents pitch and the x axis represents the forward movement of time. The thick horizontal bars represent the specific pitch and its duration against time, with the reference pitch established by the position of the x axis as it intersects the y axis. Such a representation is fully consistent with our definition of music as sound organized against time.

Until about 1000 C.E., music notation in Western Europe was a mnemonic system, in principle like the one used by the Greeks and in other cultures. The founder of modern music notation was Guido d'Arezzo, who invented a system very reminiscent of the Cartesian coordinate system above. He used four parallel lines to represent the reference pitches and various symbols inscribed on this **staff** (or **stave** in British English) to represent how the pitches should be sung . He also invented our Do-Re-Mi-Fa-Sol-La-Ti-Do system as a mnemonic for remembering the pitches of the diatonic scale. His method of notation was read left to right, going forward in time.

Now, music could be written down and then read and reproduced by another musician who had never heard the music before. This was an immense advance and no other culture had devised such a system. Today, we routinely use two 5-line staffs, one for the **treble clef** where the defining reference pitch is G above middle C and the other for the **bass clef** where the defining reference pitch is F below middle C. Middle C is produced with a white key in the middle of the piano keyboard, usually just slightly to the left of the nameplate above the keyboard. We also commonly use other clefs such as the **tenor** or **alto** clefs where the reference pitch is designated as middle C.

Let's now turn to the piano to illustrate these concepts.[4] The goal of this book is **not** to teach you the fine points of reading Western musical notation, though you may well have been exposed to notation in the past and, if so, much of this is already very familiar. But if you have access to any keyboard, whether a simple toy, a grand piano, or perhaps the easiest option for many, **a virtual on-line piano** that works through the speakers of your computer, you can play these concepts for yourself without any knowledge of notation.

The Piano Keyboard

The piano keyboard is organized elegantly into white and black keys.

Because of convention, we musicians label the white keys by the letters of the alphabet, A, B, C, D, E, F, G and then start again with A. Each piano key labeled A produces a pitch that is an octave above the nearest A to its left and an octave below the nearest A to its right. This is the case with each of the letters. The diatonic scale of Pythagoras can be played on the white keys of the piano, starting with C and proceeding to D, E, F, G, A, B, and then to the octave of the starting pitch, C'. If you do this on the piano, you will discover that you play the most familiar scale in Western music, the C-major scale. This is Guido d'Arezzo's Do-Re-Mi-Fa-Sol-La-Ti-Do. Note that the pentatonic scale is easily played on the white keys as well: C – D – F – G – A – C'.

One bit of nomenclature about the diatonic scale is important here. The starting note, in this case, C, is called the **tonic** and establishes the **key** (the tonal center) of the scale. The perfect 5th, in this case, G, is called the **dominant** and the perfect 4th, in this case, F, is called the **sub-dominant**. All pitches in the scale are named after their position in the scale. D is the second, E, the third, etc. Summarizing for the key of C-Major:

C = the tonic, the first pitch of the scale

D = the second, the second pitch of the scale

E = the third, the third pitch of the scale

F = the fourth, also known as the "perfect" fourth, or the subdominant

G = the fifth, also known as the "perfect" fifth, or the dominant

A = the sixth, the sixth pitch of the scale

B = the seventh, the seventh pitch of the scale

C'= the tonic, an octave higher than the original tonic

At this point, you can undoubtedly figure out how the word "octave" comes to us. The Pythagorean diatonic scale has eight pitches, with the first and eighth being octaves of each other. The Latin for eight is *octo* and hence, we have octave.

What about the black keys of the piano? It turns out that the diatonic scale is not the most basic convention within contemporary Western music. Instead, it's the chromatic scale, which divides the octave into twelve equal intervals. As an aside, I should point out that the 12-interval chromatic scale is by no means universal across all cultures, even though it has influenced most of the world's music. For example, the classical music of India employs 22 intervals to divide the octave, providing many additional musical possibilities. Much Middle Eastern music divides the octave into 24 intervals. Still, despite the greater number of intervals, musicians tend to use only a subset of these pitches for any particular composition, utilizing the remaining pitches for ornamentation of the melodic line.

The black keys in the piano keyboard diagram below are labeled with the pitches conventionally assigned to them. These symbols are easily decoded. The black key to the right of the white C key, for example, is labeled C-sharp or D-flat. That black key of the piano sounds a pitch one **semitone** higher than C but one semitone lower than D. A semitone is defined as one-twelfth of the octave and is also known as a **half step**, with a **whole step** (two semitones) between C and D, of course. You'll immediately realize that there is no black key between E and F or between B and C. Those pitches are already one semitone apart. If you count the individual pitches represented, you'll find that 12 intervals separate C to C'. We'll use the prime (') symbol from this point forward to indicate an octave (') or two octaves (") or three octaves (''') above the starting point.

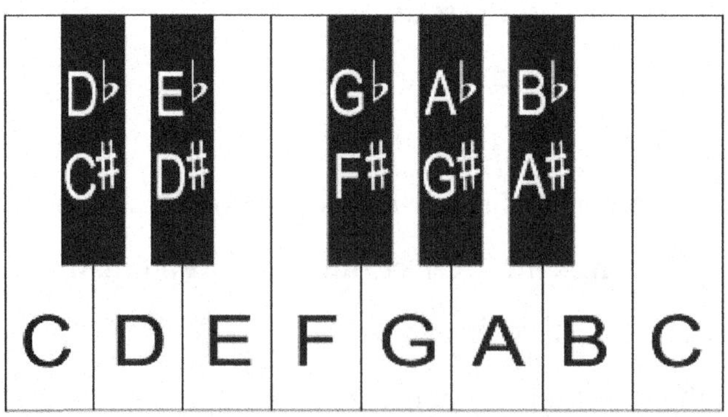

Temperament

The chromatic scale on the modern piano is **equally tempered**, which means that the frequency of each semitone is the 12th root of 2 (also represented as or $2^{1/12}$ or 1.05946309) times higher than the one to the left of it. From the standpoint of physics, such temperament makes each note on the piano slightly out of tune compared with the Pythagorean standard of perfection.

A derivative of Pythagorean thought is a scale tuned with **just intonation**. Recall from our discussion above that Pythagoras was quite enamored with the whole number ratios that defined his ideal scale and while just intonation is not identical to the Pythagorean system, it certainly isn't equally tempered. Just intonation is based upon the natural harmonic series and is determined by simple numerical ratios of the frequencies of one pitch to another.[6]

While some people very much dislike the compromise that results from dividing the octave into 12 **equal** intervals, the advantage of equal temperament is enormous. First, the diatonic scale can now start on **any** pitch of the entire chromatic scale and still sound like the same diatonic scale. From the standpoint of music theory, the scale can therefore be played in any **key**. Second, the 100 musicians of an orchestra can now perform together in any key, regardless of whether they are performing on a variable pitched stringed instrument or a mostly fixed pitch wind instrument[5] or an entirely fixed pitch piano.

Equal temperament enables a large assembly of musicians, playing very different types of instruments, to perform together. Without equal temperament and an octave divided into equal intervals, this would be impossible. Hence, Middle Eastern music and the classical music of many other areas of the globe is usually made with much smaller **ensembles** (groups of musicians). Such non-Western music also emphasizes melodic invention rather than underlying harmony, for the same reason. We'll discuss this at greater length in Chapter 13.

Here's a chart of the frequencies in Hz of the chromatic scale in just intonation and equal temperament, where C4 is middle C on the piano:

Note	Just Intonation (Hz)	Equal Temperament (Hz)	Difference (Hz)
C4	261.63	261.63	0.00
C4#	272.54	277.18	+4.64
D4	294.33	293.66	-0.67
E4♭	313.96	311.13	-2.84

E4	327.03	329.63	+2.60
F4	348.83	349.23	+0.40
F4#	367.92	369.99	+2.07
G4	392.44	392.00	-0.44
A4♭	418.60	415.30	-3.30
A4	436.05	440.00	+3.94
B4♭	470.93	466.16	-4.77
B4	490.55	493.88	+3.33
C5	523.25	523.25	0.00

To explain the calculation, the frequency of C4, middle C on the piano, is 261.63 Hz. The C# to its right, in the just system, would have a frequency of 272.54 Hz. In equal temperament, however, the pitch is defined as (261.63) x (1.05946309) = 277.18 Hz.

If we illustrate the same concept with the "perfect" fifth, the G, which in the just system should be the "beautiful" and exact 3/2 ratio higher than the tonic, making it 392.44 Hz, the now equally tempered G is no longer quite so "perfect" because it is reduced in frequency to 392.00 Hz. While the difference is small, it is audible, especially with harmony. Hermann von Helmholtz, a 19th century physicist interested in acoustics, referred to this quality as "restless" or "rough" and found it particularly offensive on the equally tempered organ.[7]

You'll note that only the octaves (the C's in this case) are exact integer ratios (2/1). This is clearly a compromise between ideality and practicality, one that some musical cultures are not willing to make. Interestingly, when musicians either sing *a cappella* (without instrumental accompaniment) or use only variable pitch stringed instruments such as violins, violas, and celli, they will often gravitate to just intonation, simply because the harmonies sound better. Vocal groups sometimes describe this move to "true" pitches as "singing the chord" and letting it "ring" as each singer adjusts pitch to work perfectly with the others, regardless of the pitch that an equally tempered piano might produce. Another metaphor is that equal temperament causes the sound always to be slightly "out of focus" and when the *a cappella* group moves to just intonation, the sound comes "into focus" for them.

Finally, we should observe that Western diatonic scales fall into two broad categories: **major and minor.**[8] The usual psychological interpretation of these scales is that the major scale seems happy and bright whereas the minor scale seems sad and dark. We'll discuss some aspects of this perception in later chapters. In addition, the minor scale comes in three forms: natural, harmonic, and melodic. Each form has slightly different conventions.

Summarizing the difference between the major and natural minor scales, both composed of 8 pitches:

C - Major: C – D – E – F – G – A – B – C' or tonic + whole step + whole step + half step + whole step + whole step + whole step + half step (octave).

Related Natural Minor (A minor): A – B – C – D –E – F – G – A' or tonic + whole step + half step + whole step + whole step + half step + whole step + whole step (octave).

Now that we have our tools in place, including our on-line virtual piano and our new-found knowledge of the scales most often employed in Western music, let's move on to an inquiry about melody itself.

Melody. At Last!

First off, let's try to define a melody.

["It's about time that we got to melody!" you are undoubtedly thinking. Applaud here if you want. And thank you for wading through the music theory with me. It will pay dividends throughout the rest of this book.]

We all have some notion of melody. Usually we rely upon our own emotional response to a sequence of tones. But a melody doesn't have to create this emotional response—at least not in everyone who hears it—for it to be a perfectly good melody.

A more general definition of melody is a sequence of tones perceived as an entity, a unit, something that "hangs together" and is coherent. Both the pitches and the rhythms with which the pitches change one to the next are critical in defining the melodic sequence. Even the harmonic underpinning can be part of its recognition.

Melodies are frequently composed of much shorter **motifs.** These sequences of pitches in popular songs are often very "catchy," easily remembered, and are sometimes called "the hook" of the melody.

Singers often use these highly memorable motifs as a mnemonic to help them to remember pitch intervals. For example,

To Remember	Motif	Piano Notes
Major second (C to D)	"Happy Birthday"	C-C-D-C-E-D
Major third (C to E)	"Oh, When the Saints..."	C-E-F-G
Perfect fourth (C to F)	"Here Comes the Bride..."	C-F-F-F
Perfect fifth (C to G)	"Twinkle, Twinkle..."	C-C-G-G-A-A-G
Major sixth (C to A)	"NBC Chimes"	C-A-F
Major seventh (C to B)	"Bali Hai"	C-C'-B
Octave (C to C')	"Over the Rainbow"	C-C'-B-A-B-C-D-C-B-A

One very important characteristic of melodies in **tonal music**, that is, music based upon the diatonic scale, is a tendency to return to the tonic (the first note of the scale) at the end of the melody or melodic phrase. This return to the tonic is nearly invariant because it allows resolution of any sonic tension that has been created in the rest of the melody and signals to the listener that this is the end. Indeed, we might say that it defines tonal music, as opposed to completely chromatic, atonal, or polytonal music. We'll discuss this much more in Chapters 10 and 12, where we'll examine the psychology of music.

Example: *Twinkle, Twinkle, Little Star*: C-C-G-G-A-A-G; F-F-E-E-D-D-C; G-G-F-F-E-E-D; G-G-F-F-E-E-D; C-C-G-G-A-A-G; F-F-E-E-D-D-C (the tonic).

This familiar little tune shows several common characteristics of many melodies in a tonal system. The overall structure is what we could call 1 – 2 – 1, or A – B - A:

Sequence 1: **C**-C-G-G-A-A-G; F-F-E-E-D-D-**C**

Sequence 2: G-G-F-F-E-E-D; G-G-F-F-E-E-D

Sequence 1: **C**-C-G-G-A-A-G; F-F-E-E-D-D-**C**

It also shows two common elements of most melodies: **repetition and contrast**. Sequence 1 is found at both the beginning and end of the tune, whereas Sequence 2 acts as a contrasting element. Psychologically, pure repetition becomes very boring, too predictable, and the listener fervently desires some novelty. Yet, once the thirst for novelty is slaked, the listener longs for a return to the familiar, and the familiar is repeated.

Good melodies, at least for popular songs, are easily sung. They rarely make a leap in pitch greater than an octave. Rather, much more often, they rely primarily upon movement to tones adjacent to each other in the diatonic scale. By the same token, memorable melodies rarely use long successions of the same pitch. Chromatic tones that are outside the specific key of the diatonic scale are very judiciously employed in such melodies, though they are frequently employed in the melodies of Western classical music. Finally, the melody has a contour—a beginning, a middle, and an end.

Part of this contour may well be a point of maximum tension brought about by the highest pitched note of the melody, with subsequent resolution of the tension, and downward movement of the melody, perceived by the listener as emotional release.

While these are good general rules, composers often violate them because the goal especially in popular music is to create a specific emotional response in the listener—happiness, sadness, nostalgia, joy, religiosity, awe, laughter, inspiration, etc. One critical factor for determining whether a popular melody becomes popular is whether it actually creates a strong emotional response, or at least pleasure in listeners, and is therefore memorable. Obviously, such melodies are far more likely to make their composer wealthy and are prized in part for that reason. The general rules that we've outlined here are helpful but, ultimately, it's the imagination, creativity, skill, and emotional sensitivity of the composer that count for everything. Arguably, melodic composition remains an art with too many variables ever to be completely predictable.

Let me cite an example. The tritone interval (3 whole steps), say from C to F#, also known as the augmented fourth, is considered to be a dissonant interval, so much so that it's been labeled "the devil's tritone." Yet, Leonard Bernstein in composing his masterpiece *West Side Story* begins the famous song "Maria" with the tritone, but then afterwards, follows the ordinary rules of songwriting quite closely.

C-F#-G, C-F#-G-A-F#-G-A-F#-G, G-F#-E-D-C-D, G-F#-E-D-C-D-C-E

It would seem that rules for composition are made only to be broken—but broken with taste and skill. Bear in mind that the surprise, and often the pleasure of music, occur with breaking rules—violating our expectations—which arguably is the case in life as well.

A Summary

1. The octave is defined as two pitches with frequencies separated by a factor of 2, that is, one twice the frequency of the other;
2. Scales are created by dividing the octave into a series of consecutive intervals;
3. The oldest commonly used division of the octave is called the pentatonic scale and relies upon whole number ratio multiples of the initial pitch, a phenomenon that the Greek philosopher and mathematician Pythagoras found beautiful and reflective of the fundamental character of Nature;
4. Based only on the octave and the fifth, Pythagoras developed a new scale called the diatonic scale, with 7 intervals and 8 pitches including the octave;
5. In Western music, the diatonic scale is a subset of the 12 intervals of the chromatic scale, which is now equally tempered so that the difference between each of the 12 is precisely a multiple of the $\sqrt[12]{2}$;
6. Equal temperament compromises the lovely, perfect integer ratios of Pythagoras but allows us to create melodies in any key and enables hundreds of musicians, as they perform with many different types of instruments, to make music together;
7. The scales that we've discussed using the notes of the piano, starting with C:
    a. Pentatonic: C – D – F – G – A – C'
    b. Diatonic (major): C – D – E – F – G – A – B – C'
    c. Diatonic (natural minor): C – D – E♭ - F - G - A♭ - B♭ - C'
    d. Chromatic: C – C# - D – D# - E – F – F# – G – G# - A – A# - B - C'
8. Melodies are formed by selecting pitches from a given scale, which in Western music is most often the diatonic scale, either in its "major" form or in its "minor" form.
9. One important concept about Western tonal music is that melodies move toward the tonic as the point of resolution.
10. Melodies are characterized by repetition and contrast and by a contour, which together provide a sense of organization and meaning.

Now, let's move on to an inquiry about the next organizing principle of contemporary Western music: harmony.

CHAPTER 6

Harmony

"Music, to create harmony, must investigate discord."
Plutarch
Biographer, Essayist

I remember being seven years old and my teacher in second grade introduced the class to a bit of singing. She decided to tell us about the idea of the **round** and began with the old standard of "Row, row, row your boat / Gently down the stream / Merrily, merrily, merrily, merrily / Life is but a dream." First, the boys in the class started from the beginning of the tune, and then the girls started from the beginning once the boys had reached their first "merrily." I remember being quite astonished that the same tune could blend pleasantly even when offset like that.

Bach

While my teacher made her description no more complex than calling this a round (it's actually a form of the classical **canon**) and may have introduced us to another old favorite, *Frere Jacques*, we were too young to understand that we'd just experienced at least the skeleton of an early form of music called **polyphony** (pronounced: puh – LIF – uh – nee). European music in the Middle Ages and into the Renaissance emphasized the singing of simultaneous melodic lines. The sonic interactions between these melodic lines could be quite complex. Composer Johann Sebastian Bach (1685 – 1750) brought instrumental polyphony to a peak of its beauty in the early 18th century, with special emphasis on the **fugue**, a musical form usually with four parts and a melody that appears in all four voices throughout the composition. The key to understanding polyphony is that melody is the predominant organizing principle of this form of music.[1]

Of course, within these compositions, different pitches are obviously sung or played together. Strict rules were employed, however. The rules arose from ideas that certain pitches when played simultaneously were pleasant but others were not.

The Basis of Harmony

Let's consider this concept together. If you use your piano or virtual piano, you can test this easily. Here is what many people would find when they sound two pitches simultaneously, though you should do the experiment for yourself:

Consonant Intervals	**Dissonant Intervals**
Octave (C + C')	Minor second (C + C#)
"Perfect" Fourth (C + F)	Major second (C + D)
"Perfect" Fifth (C + G)	Tritone (C + F#) *(the devil's tritone)*
Sixth (C + A)	Minor 7th (C + B-flat)
Major Third (C + E)	Major 7th (C + B)

While the perceived pleasant or unpleasant character of two pitches sounded together reveals something about musical harmony, it's really only the beginning of the story. Let's consider an extremely brief history of harmony in Western music to gain some more insight.

We've already observed that the difference between the voices of men and women (and prepubescent children) is approximately an octave. As our ancient ancestors began to learn about singing, this octave difference was likely their first interval of harmony because it was based upon a physical difference between men and women or children.

With time, as musical sensibility increased, it's quite likely that humans began to sing in fourths and fifths, as well as the octave. We've already speculated that the pentatonic scale was probably the earliest scale invented—possibly nearly universally across human cultures—and the perfect fourth and perfect fifth intervals are very important in that scale. While music created solely with harmony from octaves (C + C'), fourths (C + F), and fifths (C + G) may seem rather boring to our contemporary ears, our ancestors undoubtedly enjoyed it greatly.

Toward the end of the medieval period, and definitely by the beginning of the 17th century in Europe, musicians were adding the third pitch of the diatonic scale to the combination of tonic and fifth. With that addition, the **triad** was born. One form of triad is C + E + G or the **C-major chord.**[2]

The Triad as the Fundamental Unit of Western Musical Harmony

The triad exists in four primary forms (here voiced with the tonic as the root), together with a number of secondary forms which omit the third and substitute a second or fourth. These include the major triad, the minor triad, the diminished triad, and the augmented triad. Here are examples with C as the root for the sake of simplicity:

C - major: C + E + G *or* 0 + 4 + 3 semitones *or* major 3rd + minor 3rd

C - minor: C + E♭ + G *or* 0 + 3 + 4 semitones *or* minor 3rd + major 3rd

C - diminished: C + E♭ + G♭ *or* 0 + 3 + 3 semitones *or* minor 3rd + minor 3rd

C - augmented: C + E + G# *or* 0 + 4 + 4 semitones *or* major 3rd + major 3rd

Western music began to change from polyphony to **homophony** (pronounced: huh – MOF – uh -nee). Rather than an emphasis on multiple simultaneous melodic lines, which is also referred to as **counterpoint**, homophony uses a single primary melody undergirded by harmony, that is, by **chords** (two or more pitches sounded together). In fact, some composers start with a **chord progression** (a sequence of chords in time) that they like and then write a melody that makes sense for that particular chord progression. There, harmony drives melody. Thus, in homophonic music, harmony is a fundamental organizing principle, at least equal to melody.

It's important to mention that musicians are **not** in any way limited to three pitches in a chord. Chords can contain many pitches, but ultimately the analysis of the chord in Western music returns to an analysis of the triad on which the chord is based. Moreover, the nomenclature of chords, that is, the conventions for naming and identifying chords, is also based upon this triadic structure.

From the standpoint of the physics of sound, the quality of each of these triadic chords can be understood, at least partially, in terms of the theory of sound waves. Let's examine the C-major chord as an example. The C-major chord, or any major chord, sounds very harmonious to our ears today. Moreover, our brains interpret the chord as a unit, with all of the pitches blending together so that all major chords in any key (that is, with any chromatic pitch as the root or tonic of the chord) have the same sonic and musical character and are recognized by our brains as a major chord.

While the blending of these pitches creates a new waveform, musicians can tease apart the individual pitches and sing them. A musician who has absolute (perfect) pitch can identify all the notes within the chord. Additionally, the chord has a distinct character that results from the blend of pitches but it does not have its own frequency, unlike each of the pitches within the chord. The chord's waveform is the summation of all of the pitches and their individual harmonics. Clearly, such a waveform can be very complex, especially if the chord has, for example, eight pitches.

Scientific Explanations for Consonance and Dissonance

Can we attempt to explain the harmonious quality of a major triad? Yes. Think about the harmonic series. Imagine three ideal strings, tuned to fundamentals of C', E', or G'. They are plucked simultaneously to produce a C-major triad. Each string, of course, produces its own harmonic series. Those harmonics interact with each other in ways that we might describe as consonant or dissonant. Let's analyze the harmonics produced by these three strings:

	Harmonic					
	1st	2nd	3rd	4th	5th	6th
C'-string	C'	C"	G"	C"'	E"'	G"'
E'-string	E'	E"	B"	E"'	G#"'	B"'
G'-string	G'	G"	D"	G"'	B"'	D"'

By inspecting this chart, the overlap between the harmonics of these three ideal strings is readily apparent. Of course, with ideal strings, we are using just intonation. Equal temperament would exacerbate the tension resulting from any mismatch of the upper harmonics. In fact, a distinct difference can easily be heard between C-major chords sounded in just intonation as compared with equal temperament, apparently in part because of these interactions in the harmonic series.

The idea that consonance or dissonance results from the interaction between the harmonics of a vibration rather than arising solely as a property of the interactions between the fundamental frequencies comes to us from the work of German physicist Hermann von Helmholtz[3], who did extensive research on the physics of sound and musical pitch in the mid-19th century. Some evidence supports his theory and it may explain part of the physical basis of consonance versus dissonance. A mismatch in the peaks and troughs of the sound waves of harmonics does lead to the "beatings" that we discussed in Chapter 4, and this in turn would lead to a perception of what Helmholtz referred to as "restlessness" or roughness in the sound, and that would presumably be described as dissonance. Clearly, the presence of "beatings" is not pleasant and there is evidence that even macaque monkeys may well perceive such mismatch in the same part of the brain as humans.[4]

In recent years, using sophisticated tools that we will discuss in Chapter 9, neuroscientists have begun to study the perception of consonance and dissonance in a different way. These teams of scientists have obtained evidence that our perception is apparently not just about the "roughness" of physically mismatched waveforms. Rather, it's also about the biology of our brain.[5,6]

First, certain people, called amusics, claim no preference for consonance over dissonance. But they can still identify the beatings of dissonant tones. Moreover, when a tone is played in one ear of a musical human subject and a consonant or dissonant tone is played in the other ear so that there can exist no physical beatings, a perception of consonance or dissonance is nonetheless reported by the person. Finally, as we shall learn in Chapter 13, the perception of consonance or dissonance seems culturally determined, at least in part. All of this indicates that a simple mismatch (technically, an alternating constructive and destructive interference) of waveforms is apparently not the complete explanation for this phenomenon. It may be important to distinguish between consonance / dissonance in a musical context versus the psychoacoustic analysis of isolated sounds.[4]

Chord Progressions

Related to this observation, we still have little understanding about how certain chord progressions can be so emotionally evocative for us. The choice of such chord progressions in a musical composition requires as much art as is required for finding wonderful melodies. We do have some rules (based upon experience and taste) that we'll learn about in a few paragraphs, but these are rules made to be broken, just as are the rules for composing melodies. We know only a little about the processing of chord progressions in the neuronal circuitry of the brain (Chapters 12 and 13). Harmony remains a delightful mystery and our scientific explanations are currently incomplete. We'll explore this mystery from neurological and psychological perspectives in later chapters of this book.

Over the last few centuries, composers have become progressively bolder and experimental in finding new chord progressions. Jazz musicians, in particular, use a very broad and rich palette of chords. Many chords, which would presumably have sounded horribly dissonant to a musician living in the 17th century, are routinely employed today. Musicians and listeners have embraced new harmonic forms and most popular music today would sound dull and uninteresting without "jazz" chords not found in earlier music. Imagine a musician who used exclusively medieval octaves, perfect fourths, and perfect fifths as the harmonic underpinning of melody. Such a musician is unlikely to gain sustained popularity in contemporary music culture.

Even so, certain forms of popular music do use a very limited number of chords. Rock 'n' roll is an example. Much such music can be performed with only three chords, as garage band musicians quickly discover. Here, of course, the primary organizing principle of the music is rhythm—not necessarily complex rhythm, but certainly predominant rhythm. Some theorists view rock 'n' roll as low information music and contrast it with jazz and Western classical music, which are far more complex, designating those forms as high information music. Each type of music has its place, however, and good, original rock 'n' roll is much appreciated by many.

Let's begin a more detailed inquiry about various chord progressions. Let's also begin empirically—very appropriate in a book about science, right? Go to your piano, real or virtual, and play the following chord (left to right) in the middle of the piano:

G + B + D + F

Musicians refer to this chord as a G7 chord because its basis is the G-major triad (G + B + D) but it adds the flatted seventh of the G-major diatonic scale (F, where F# would be the ordinary 7th pitch of the scale).

Play the chord repeatedly. How do you feel about this chord? Do you feel perfectly fine with it or is it somehow "just not right?" Perhaps it's incomplete?

Most people report a sense of incompleteness, even anxiety or uneasiness, when this chord is played repeatedly on the piano. One of my music teachers (perhaps a bit sadistically) delighted himself by entering a room that was furnished with a piano, sounding a G7, and then walking out of the room, leaving everyone else to dwell in the unease, until someone went over to the piano and **resolved** the chord, as we will below.

So, now play another chord:

G + C + E + G'

Many people react with a long sigh of relief. Ahhhhhhhhh!!!

This is a C-major chord with the dominant (the fifth, G) as the lowest note, a very common **voicing** (choice of the order of pitches) of a major chord. So, what happened?

G → G (stayed the same)

B → C (up one-half step)

D → E (up one step)

F → G' (up one step)

To our ears and brains, the G7 chord has within it the kernel of the C-major chord. It is a **leading chord**, one that feels **unstable** and requires **resolution**. That's what we did. We **resolved** the chord. And, thus, we relieved the tension.

We've just demonstrated a fundamental principle of emotionally evocative chord progressions:

1. Establish an expectation;
2. Create a tension related to that expectation;
3. Resolve the tension by fulfilling the expectation. Or sometimes unexpectedly delay the resolution.

How a combination of pitches can create the pleasure of an expectation / tension / resolution sequence in the neuronal circuits of our brains remains a mystery, which we'll investigate a bit in later chapters of this book looking for some measure of insight. It is, however, a wonderful part of music. A progression of chords, at least as much as a sequence of melodic intervals, has enormous power to create and to resolve tension. Most of us find such experience highly pleasurable.

What are some commonly used chord progressions? To make that inquiry, it will be useful to discuss the rudiments of how musicians name chords employing a shorthand system based upon the notes of the diatonic scale. Here's an example for the C-major scale:

Triad	Chord Name	Shorthand	Alternative System
C+E+G	C major	I	C or C_{maj}
D+F+A	D minor	ii	D_{min}
E+G+B	E minor	iii	E_{min}
F+A+C'	F major	IV	F or F_{maj}
G+B+D'	G major	V	G or G_{maj}
A+C'+E'	A minor	vi	A_{min}
B+D'+F'	B diminished	vii	B_{dim}

The rationale for these systems is quickly apparent. With the shorthand system, the Roman numeral refers to the pitch position on the diatonic scale. Upper-case numerals indicate that the triad is major and lower-case numerals indicate that it is minor. This system has the advantage that it applies to all keys, meaning that regardless of the tonic, the system still works. The alternative system often used by jazz musicians, guitarists, and in fake books (books with the melodies written out but with only **chord symbols** to indicate the harmony) is self-explanatory, versatile, and quite precise.

While the type of triad is the basic description used for Western chords, the voicing of a chord can vary widely. For example, the 3rd can be placed at the lowest position. This is called the **first inversion** of the chord, or the 5th can be placed at the lowest position, making the **second inversion**. Additional notes can be added to the triad, especially as an octave of the lowest pitch. One common addition is the flatted 7th, to produce the 7th chord, as we observed in our G7 example above.

Here's a chart for us to explore some common chord progressions:

Chord Progression	Comment
V7 – I	see our experiment above
I – IV – V – I	perhaps the most common progression of Western music
I-I-I-I-IV-IV-I-I-V-IV-I-I	12 bar blues
I-vi-ii-V-I	doo-wop progression
I-IV-vii-iii-vi-ii-V-I	circle progression

Some things to think about:

1. In tonal music, chord progressions establish the key and drive ultimately back to the tonic (I) for resolution;
2. Chords can be modified in many different ways through the addition of pitches, especially 7ths, 9ths, and 11ths, as well as through the choice of the primary triad (major, minor, augmented, diminished, etc.), deliberate contrast with the base line, and choice of voicing of the chord, including the octave in which the pitches occur;
3. While common progressions usually prevail in popular songs, jazz musicians use these merely as a starting point;
4. Composers often establish a bass line and melody first, creating the chord progression from the bass line alone;

5. Chord progressions can easily be formed from any of the types of minor scales, not just the major scale;
6. Composers will often change keys, that is, **modulate** using chord progressions, to obtain an emotional response from the listener;
7. While composers occasionally use "block" chords, where the chord is sustained beneath the melody, more often the notes of the chord are separated and given rhythmic interest and propulsion.

This chapter can only scratch the surface of the complexities of harmony. But we've learned a few things. My experience is that listening to *a cappella* singers, especially male choirs, where the range of pitches is not as wide as with mixed choirs, is an effective way to dissect harmonies. Barbershop quartets, with only four voices and a limited number of chords, are especially valuable in this regard, as are small groups of male singers, up to perhaps eight vocalists, because the chords can obviously then be a bit more complex.[7]

Rhythm, melody, and harmony are the ordinary organizing principles of contemporary Western music. Each plays a vital role. Together, they enrich our lives in extraordinary ways.

Our next question, though, is about the devices—the instruments—that we use to produce these emotionally evocative sounds. How do they work and how does science help to make them possible in their modern forms?

CHAPTER 7

Making Music

"It's easy to play any musical instrument: all you have to do is touch the right key at the right time and the instrument will play itself."

Johann Sebastian Bach
Composer

"A gentleman is someone who can play the accordion, but doesn't."

Tom Waits
Singer, Composer, Actor

My first wind instrument was the Tonette, a small, plastic, tunable, end-blown song flute (a bit like a recorder) used often in the 1950's to provide elementary school children with their first experience in instrumental music and to teach them the basic elements of Western musical notation. Tonettes were inexpensive and indestructible—two attributes that recommended them for elementary level education. At eight years old and in the fourth grade, I remember the wonder of discovering that I could place my fingers over the correct tone holes of the Tonette, blow into the simple mouthpiece, produce pitches, and play a real tune. Frankly, I was then hooked on making music with instruments other than the piano.

The science behind the construction of that very simple musical instrument used by children is the same science that we'll discuss here as we learn about the instruments of the modern symphony orchestra. The theories developed by physicists have been helpful to those who design and build musical instruments as well as to those involved in the design and construction of concert halls. Throughout modern human history, music has inspired discoveries by physicists and mathematicians, certainly from the time of Pythagoras. In addition, biologically derived materials have performed a key role in the fabrication of musical instruments, as we shall see, and therefore biologists from several different disciplines have been directly or indirectly involved with improving musical instruments. A continuous "cross-pollination" seems to have existed between the contributions of musicians and scientists over the centuries.

Let's do a **thought experiment** together. In German, we'd call this a *Gedanken experiment.* The question is: based on our observations of Nature and our knowledge of the physics of sound, what is required to fabricate a musical instrument and are there common principles in the design of all musical instruments?

Vibrators, Drivers, and Resonators[1,2]

Thinking back to Chapter 4, we'd instantly say that all musical instruments require something that vibrates. So far, we've mostly discussed strings, but perhaps other things could vibrate as well. Regardless of what it might be, something has to vibrate. Let's call it the **vibrator**. We could call it an **oscillator**, though vibrator is a more specific term for our purpose here.

If something vibrates, something else must set the vibration in motion. Even casual observation indicates that we don't observe musical vibrations happening spontaneously without cause, that is, without some energy input to get the vibration started. Let's call whatever initiates the vibration the **driver**.

Finally, if we think about a vibrating string all by itself, fixed at two points and suspended in air, we know that it produces sound of low amplitude—it's just not very loud. (Try stretching a rubber band between your fingers and plucking it to get an idea of this.) The string has very little surface area and therefore doesn't transfer its vibrational energy to very many air molecules. Yet, we observe that musical instruments, including the human voice, can produce extremely loud sounds. Just think of a playground filled with screaming children. So, based upon those observations, musical instruments must have something that increases the amplitude of the sound wave, and it may even shape the waveform by reinforcing certain harmonics over others. Let's call that the **resonator**.

The phenomenon of **resonance** is a straightforward concept. An example known to everyone is the opera singer with a powerful voice who is able to sing at precisely the resonance frequency of a wine glass, causing it to vibrate sympathetically, so much so that the vibration stresses the glass and causes it to shatter. All resonators have preferred frequencies at which they vibrate, and as the resonator vibrates, it adds to the amplitude of the original wave, thereby amplifying the sound. The design of resonators is a combination of art and science, but they are a critical component of musical instruments, as we shall see.

Perhaps we could examine a very simple, one-string instrument called the **Canjo**.[3] This instrument can be purchased from several different sources, either finished or as a kit. It demonstrates all of the concepts that we've been discussing here. First, it has a single steel string, drawn taught between two posts, which are attached to a wooden fingerboard. The string can be tightened or loosened for tuning using a simple tuning peg at one end of the string. At the other end, the string is affixed to a tin can. The fingerboard has frets (ridges) to assist the musician in finding the string length that gives rise to the desired pitch. The instrument can produce the major diatonic scale in two keys, and can quickly be tuned to a fundamental pitch of middle C. The salient points:

1. The driver here is ultimately human muscle (as it is in many, though not all, musical instruments), in this case acting through a finger or pic to pluck the string;
2. The vibrator is the steel string;
3. The resonator is the tin can. The tin can amplifies the sound from the vibrations of the string by vibrating sympathetically with the string, resonating more with certain harmonics and even adding its own preferred harmonics, resulting in the plucked string sounding like, well, a canjo, instead of say, a guitar, even though the same type of string would be used on a guitar.

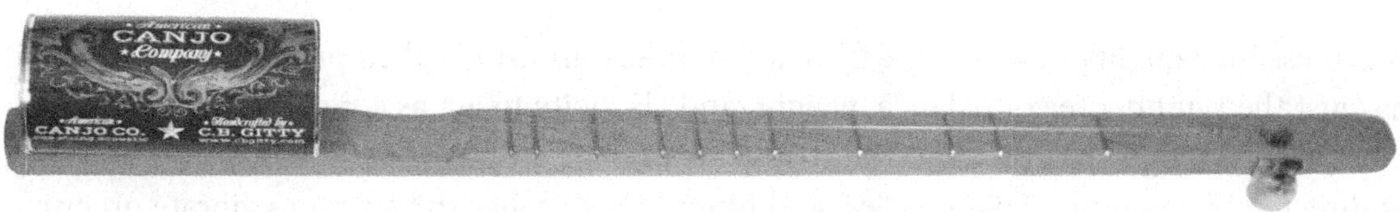

Our initial thinking certainly seems to make sense with regard to this simple instrument. But surely, contemporary real musical instruments are more complex. To gather evidence, we could assemble all contemporary musical instruments and study them. Or, we could go to a concert. Which would be more fun?

The String Family

If we attended a concert by a large symphony orchestra, the first thing we'd notice was a veritable army of musicians in front who were playing on 4-stringed instruments of various sizes: violins, violas, celli, and double basses (small to large). We'd notice that the smallest instruments sounded at the highest pitches and the largest produced the lowest. Remember that the frequency of the sound wave produced by a vibrating string is a function of its length, mass, and tension. The strings of the double basses are long and thick (greater mass) and therefore produce a lower pitch. Recall: greater length and / or greater mass and / or less tension ➔ lower frequency of vibration.

If afterwards we boldly approached the concertmaster (the principal violinist) to learn more about the string family of instruments, she'd undoubtedly show us her violin and explain its construction. The four strings, made sometimes of sheep gut, though today often of some type of metal or composite, are tuned in ascending perfect fifths: G (below middle C), D, A', and E'. Violinists, after long study, develop muscle memory about where to place the tips of their fingers on the strings to stop them—that is to alter the length of the strings—and thereby to produce pitches higher than the fundamental pitch of the particular string. Unlike the canjo (guitar, ukulele, and many other instruments with strings), violins do not have frets on their fingerboards to guide the placement of the fingers.

Our concertmaster would also point out the importance of the construction of the body of her violin and of the bow that she uses to set the strings in vibration. The top sounding board of the violin is made from quarter-sawn spruce, which is the same material used in high quality piano sound boards, the back and ribs from maple, the finger board from ebony, the neck and bridge from maple. The bridge, small though it is, forms the link between the vibrating strings and the body of the violin. Tiny differences in its placement can dramatically affect the sound.

A professional quality bow is made from a rare Brazilian wood called pernambuco, so that it has just the right degree of balance, weight, and elasticity to act as a perfect device to transfer energy through friction from the rosined horsehair stretched on the bow to the string, as directed by the musician's muscles, and ultimately to cause the string to vibrate precisely, sensitively, and beautifully.

Pernambuco (***Caelsapinia echinata)***

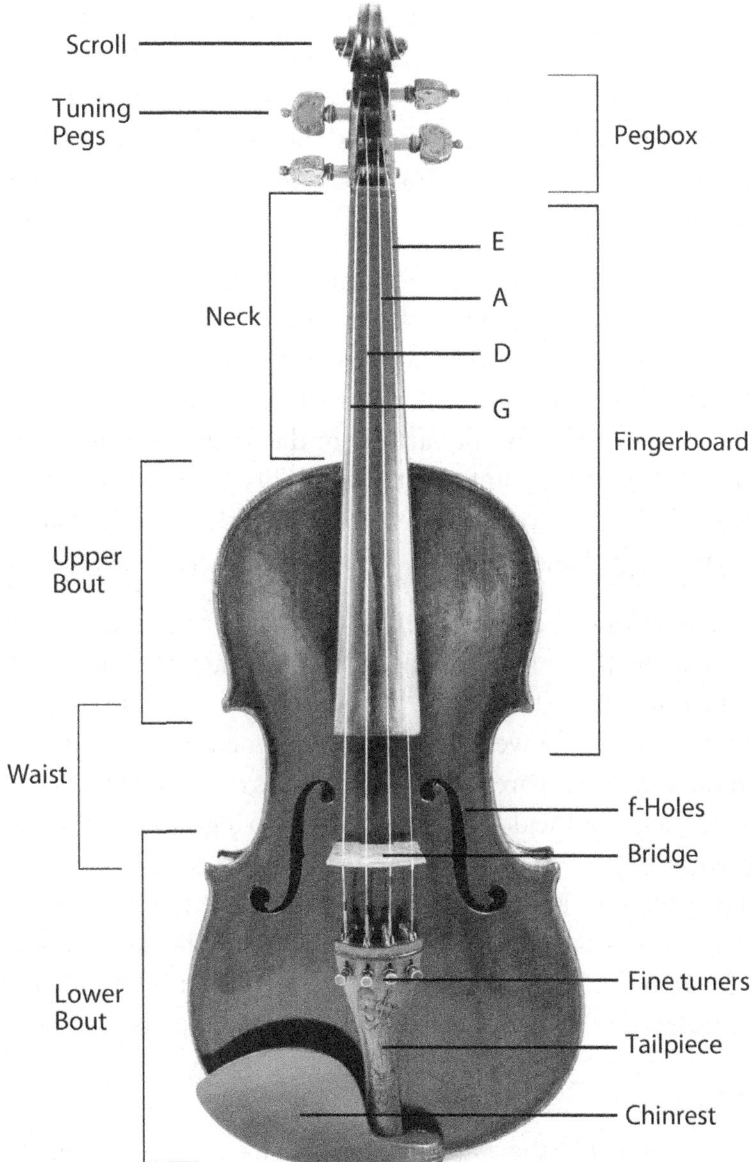
Scroll
Tuning Pegs
Pegbox
E
A
D
G
Neck
Fingerboard
Upper Bout
Waist
f-Holes
Bridge
Lower Bout
Fine tuners
Tailpiece
Chinrest

A violin bow

The quality of a violin's sound, that is, the relative amplitudes of its harmonics, depends on the skill of the musician, the resonance characteristics of the body of the violin, and the quality of the bow. The construction of the body of the instrument and of the bow require both science and art. Selection of the proper botanical materials, grown correctly, is clearly important. Pernambuco is an endangered species and therefore the ongoing work of biologists to ensure its conservation and survival is key to the future of string instruments as we play them today in the modern orchestra.

What makes a Stradivarius violin sound so beautifully? This is a matter of debate.[4] Scientists have certainly investigated the question but have come to no consensus conclusions. Some studies even show that accomplished musicians listening to various excellent violins cannot discern the Stradivarius. But violinists claim that there is something magical in performing on the Strad. Perhaps it's the wood, or the varnish, or the design. At this point, no one knows for certain. All evidence, however, points to the complexity of resonance, which is so complicated in the body of a violin that we cannot predict it completely. All we know for certain is that when it's right, and in the hands of a highly skilled musician, it's breathtakingly beautiful.

As an aside, one challenge that all musicians in an orchestra encounter is in playing synchronously with the beat set by the conductor. In addition to watching the conductor, they will sometimes watch the down-bow (the start of the note) of the concertmaster. Light waves, reflected from her violin and arm, traveling at 300 million meters per second; appear virtually instantaneously in their eyes compared with sound waves at 343 meters per second. But this presents a special challenge for the double basses. Their strings are so massive, with so much **inertia** (a term used in physics to describe the tendency to remain at rest or to remain in motion) that bass players must anticipate the start of the note to remain temporally aligned with other musicians in the orchestra. This is exacerbated by the fact that the double basses are often located far away from most of the rest of the orchestra. Physics and mathematics reign supreme, regardless of whether music makers are always conscious of them.

The Woodwinds

After gratefully bidding *adieu* to the concertmaster, we have the opportunity to speak with the principal oboist of the orchestra. Like the concertmaster, he occupies a special spot in the hierarchy of orchestral musicians. First, it's his responsibility to provide the "tuning note" for the entire orchestra. Secondly, as a matter of tradition in many orchestras, the principal oboist may be considered the first among equals in the woodwind section that is comprised of piccolo, flutes, oboes, English horn, clarinets, bass clarinet, bassoons, and contrabassoon—and very rarely, as a guest participant, the saxophone. Note that the saxophone is a recent addition to the woodwind section and often isn't quite completely accepted by so-called serious composers, though it has certainly conquered the world of pop music and jazz.

With regard to the "tuning note," the oboe is ideally suited to provide this standard pitch that acts to unify the entire orchestra. Interestingly, the "standard" pitch of the A above middle C, supposedly set at 440 Hz, is not really an immutable standard at all. Historically, A_4 (the fourth A on the piano, counting from the left) has ranged from 422 Hz (Mozart's preference) to as high as 450 Hz. The higher pitch adds brilliance to the sound of the instrumental ensemble but can cause singers to strain considerably. In 1939, an international commission established 440 Hz as the standard, but many well-known orchestras still tend to tune higher. Some European orchestras apparently tune to 444 or 445 Hz. Even if given an initial pitch of 440 Hz, musicians will often drift up in pitch (become sharper by a few Hz) over the course of a performance. Completely uniform pitch, as we learned when we discussed just intonation and equal temperament in Chapter 5, is a goal rarely achieved. It's always a balance and compromise.

So, why is the principal oboist designated to provide the tuning note for the orchestra? Some of us musicians would cheekily argue that it has to do with the dominating personality characteristics of some oboists, but the verifiable evidence for such a conclusion, while tantalizing, is insufficient. The actual reason has to do with the characteristics of the oboe as a musical instrument. To inquire about that, we must begin a discussion of woodwind instruments in general.

Within the name of this group of instruments, "wood" + "wind," exist two implied characteristics. The sound is driven by **wind** from the mouth of the performer, which is in turn driven by the diaphragm, a large muscle responsible for breathing, as well as by other accessory muscles in the abdomen and between the ribs. All of these woodwind instruments consist of a tube, historically made from wood, but now frequently made from precious metals (silver, gold, platinum) in the case of flutes, or from brass in the case of the saxophone family. Interestingly, high quality professional piccolos are still made from wood, as are oboes, English horns, clarinets, bass clarinets, bassoons, and contrabassoons. Grenadilla (African blackwood, also known as mpingo) is most often used in the manufacture of professional piccolos, oboes, English horns, clarinets, and bass clarinets, whereas sycamore maple or sugar maple is used for bassoons and contrabassoons. Whatever its composition, the tube, together with the entire airway of the performer, acts as the resonator for this group of instruments.

Bassoon

Oboe

Clarinet

Piccolo

Grenadilla (African blackwood)
Dalbergia melanoxylan

Incidentally, the conservation, preservation, and cultivation of the endangered African blackwood tree is another example of biologists diligently contributing to music-making. This tree grows very slowly, typical of those trees that produce extremely dense, heavy wood. In fact, grenadilla wood is harvested for the manufacture of musical instruments only after 70 to 200 years growth. The advantage of grenadilla and similar very hard and dense woods, such as rosewood, is in their stability, resistance to moisture, ease of machining and perhaps in the quality of the sound produced due to the vibration and resonance of the wood itself, though this is a matter of some controversy.

The Drinking Straw Oboe[5]

To begin to investigate the acoustical properties of woodwinds, let's do another simple experiment. We're going to make a **drinking straw oboe**. This is an experiment that is particularly fun to do with a younger child. Detailed pictorial instructions are easily accessible online.

Go to your kitchen and find some plastic drinking straws. Take a straw and pinch 2 or 3 cm (about an inch) of one end flat, or at least as flat as possible. Now, take scissors and cut the flattened end into a triangle about 1.5 cm (1/2 inch) long. The end of the straw thereby becomes the point of the triangle.

Take the cut triangle end of the straw, place it in your mouth, put your lips around the straw so that it's sealed, but don't apply so much pressure as to close off air flow. Blow through the straw in such a way as the triangle begins to vibrate inside your mouth. This may take a few attempts to get it right.

Before long, you'll begin to hear a buzzing from the straw. You'll find that you can adjust the pitch of this buzzing in at least two ways: (1) by changing the position and pressure of your mouth on the straw and (2) by changing the length of the straw by cutting it to shorter lengths with scissors. The shorter the straw, the higher the pitch will be. This simple drinking straw oboe demonstrates the key physical principles of woodwinds, just as the canjo demonstrated the main principles of string instruments.

So, what's going on here? First, the vibration itself is generated by differential air pressure. As you blow through the straw, air rushes past the open triangle through the straw, lowering the air pressure between the triangular flaps. Because the air pressure is less, the flaps close. But now, air from your breath no longer flows, and atmospheric pressure (from the open far end of the straw) causes the flaps to spring open. Air from your breath can flow between the flaps once again, they close, and the process continues cyclically.

This vibrator, of course, disturbs the air molecules within the straw and creates a **standing wave**[6] inside the tube of the straw. In terms of its physics, the standing wave within the straw is very similar to that of a vibrating string, which also establishes a standing wave, in that case, on the string. Instead of the wave physically residing in the string, however, the wave resides in the compression / rarefaction cycles of the air molecules **within** the straw.

Comparing the standing wave of a vibrating string with a standing wave inside a tube filled with air:

1. A shorter tube, like a shorter string, results in a higher frequency of the standing wave;
2. Like a string, if all other variables are constant, a tube half as long will produce a frequency twice as great—the octave;
3. Increased air temperature inside the tube, like a tighter string, results in a higher frequency of the standing wave;
4. While strings can vary in mass, tubes can vary in shape e.g. cylindrical, conical, and in organ pipes, even rectangular. The shape of the tube profoundly influences the fundamental frequency for that pipe, as well as its mix of harmonics;
5. For both strings and pipes, a set of harmonics is produced but while any string will produce all of the harmonics, tubes with different configurations will have very different mixtures of harmonics. More about this later.

Let's return to the real oboe. Rather than a bit of triangular plastic as in our straw, the oboist uses a specially prepared piece of reed cane from the plant *Arundo donax* L., a very tall grass (20 feet) that thrives in Mediterranean climates but is now found in temperate climates throughout the world. Like our drinking straw triangle, this is a **double reed**, that is, it has two halves that vibrate against each other. The fabrication of oboe reeds is time-consuming and demands much skill. Any given reed is unlikely physically to withstand the stresses and strains of rapid vibration much longer than 10 hours. In fact, many oboists use one reed per concert. Moreover, the quality of the cane for making top performing reeds is highly variable and has been the subject of scientific research and publications, including studies by the present author.[7] Current understanding indicates that more mature cane is likely to produce better reeds than cane that is younger and softer. Of course, cane can become too mature and too hard for use as musical reeds and therefore, as with most things, this is a compromise.

Oboe reed, with vibrating cane at the top, and cork for insertion into the oboe at bottom:

Arundo donax L. growing wild near a stream.

And now we come to the reason that the principal oboist provides the tuning note for the orchestra. During the preparation of each reed, the oboist must tune the reed and make certain that it reliably produces A440—or whichever frequency is set as the tuning note by his particular orchestra—when inserted into the top end of his oboe. This process requires considerable skill and oboists in the past have routinely matched the pitch produced by the reed to a tuning fork, but they now use electronic tuners to ensure proper frequency. Thus, the principal oboist is the keeper of the standard pitch and provides all of his fellow musicians with an accurate and reliable "A" at the beginning of a rehearsal or concert.

Each woodwind instrument employs one of three types of reeds as the vibrator. We just mentioned the double reed that relies on two halves of specially shaped reed cane beating against each other, thereby disturbing the air within the tube. Oboes, English horns, bassoons, and contrabassoons (and yes, often bagpipes) all use double reeds of various sizes and configurations.

Another category of woodwinds, the clarinets and saxophones, employs a **single reed** attached to a mouthpiece. This reed is also made from *Arundo donax* L., but its design is considerably simpler. Because single reeds are readily available for purchase commercially, many professional single reed players do not make their own reeds, whereas nearly all professional oboists and bassoonists invest this considerable effort.

Clarinet mouthpiece with single reed attached.

As an aside, chemists and engineers have attempted to create a totally synthetic reed that would match the acoustic properties of cane reeds but would be far more durable and reliable. They've had some success. A few professional clarinetists and many saxophonists use these polypropylene reeds in performance, but the vast majority of single and double reed professional musicians reject them, despite their advantages, finding that the plastic simply cannot provide the range of timbres permitted by good cane, nor can it match and blend with the sound qualities of other woodwind players in the orchestra. This debate about cane versus plastic and the effort to make a totally synthetic reed are ongoing. Musicians can be opinionated, like everyone else.

The third type of woodwind reed is not made from cane. Rather, it is simply a sharp edge across which the musician blows a stream of air. Hence, it is called an **air reed** and is found in piccolos, flutes, recorders, flue pipes in organs, my elementary school Tonette, and similar instruments. When the musician blows air across the edge, the stream is split into two streams oscillating back and forth, creating turbulence. This causes a disturbance in the air molecules within the tube and thereby generates a standing compression-rarefaction wave. In essence, the turbulence (air first going over the top, and then through the instrument) caused by splitting of the air stream creates a vibration.

Diagram of air reed showing the turbulence created in air flowing against a sharp edge.

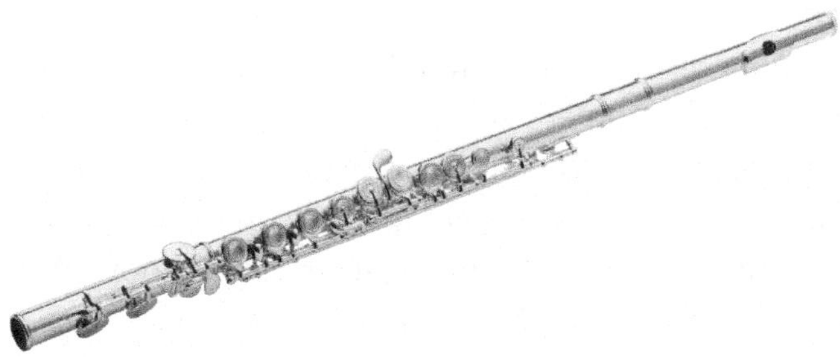

Flute with blowhole at right end in this picture

Regardless of whether the woodwind instrument uses a double, single, or air reed, the mechanics for determining pitch are very similar. The tube is replete with tone holes, precisely placed, with precise diameters, and shaped in an exacting manner, with dimensions in part derived empirically and in part derived from the laws of physics. In addition to the tone holes, the bodies of the instruments have a complex system of pads, keys, rods, and springs that allow the covering and uncovering of tone holes quickly. This system enables the musician to alter the length of the tube very rapidly to produce the desired pitch at the desired time, in a manner entirely analogous to the way a violinist shortens or lengthens a string through the placement of her fingers.

Of course, just as with the violinist, much skill is required, and many thousands of hours of practice must be invested to make this process work smoothly and accurately so that the musician produces beautiful sound. In addition, while the violinist need not worry about breath or reeds, the woodwind player is concerned about both, and this adds another element of challenge. On the other hand, because woodwind instruments produce a pitch that is relatively fixed compared with the infinite variability of a string, this part of the task is easier in the initial stages of learning to play a woodwind instrument, though precise intonation still demands much skill.

The human voice, while not exactly belonging to any category of instrument, is perhaps most like the woodwinds. Like them, the vibrator is driven by air, in turn driven by the diaphragm and accessory muscles. Instead of a cane reed, the singer uses her vocal cords, which vibrate a bit like strings, but also a bit like a reed, in that they are set in motion by moving air. Finally, the singer's entire lung cavity, sinuses, throat, and mouth are all part of her resonator, just as these are also part of the resonator for woodwind players, especially for clarinetists, saxophonists, bassoonists, and oboists, who all take the vibrating reed into their mouths.

The Brass Family

After such an enlightening discussion, we finally must say farewell to our principal oboist and move on to the principal trumpet player, who has been sitting patiently waiting for us. It's an ongoing joke among musicians that the principal trumpet must possess the strongest ego of any musician in the orchestra, save perhaps, the conductor. First off, the loudness and directionality of the trumpet sound make any mistake entirely obvious to the audience. But perhaps more challenging, composers frequently score an exposed trumpet entrance late in the composition, and therefore demand that the trumpet make a beautiful entrance without benefit of sufficient warm up. This requires not only skill but also courage and immense self-confidence.

We quickly learn from her that the trumpet belongs to a family of instruments known simply as the **brass**, and indeed, they are usually constructed of brass tubes, so all does seem quite right with the world. No surprises here. The physics of their acoustics is somewhat similar to the woodwinds. First, the driver for vibration is air, driven in turn by the diaphragm and accessory muscles. Instead of a reed, however, brass players use the buzzing vibration of their own lips into a cup mouthpiece. This buzzing disturbs the air molecules in the brass tube, creating a standing wave, just as found in the tubes of woodwinds.

Woodwind players open and close tone holes to change the length of the tube to produce different pitches, while the actual, maximum physical length of the tube of their instrument remains constant because it's fixed. In other words, a clarinet's actual length doesn't change as it's played but its functional or virtual length does. The venting of the woodwind tone hole shortens the functional length of vibrating air column and thereby raises the frequency. In contrast, brass instruments quite literally change the length of the tubing itself, either by opening or closing valves that divert the standing wave into another section of tubing that may be longer or shorter or by directly changing the length of tubing, as in a slide trombone. The valves, a relatively recent invention, permit far more rapid note changes, but a slide trombone can produce unique effects such as *glissandi.*

A few brass instruments

As with the woodwinds and strings, brasses come in a range of sizes in the orchestra. Trumpets, French horns, trombones, and tubas make up the standard brass section. The same physical principles of standing waves in a tube apply to them as for the woodwinds. A longer tube enables a lower fundamental frequency, and warmer air causes the pitch to rise in frequency. The shape of the bore, whether conical, cylindrical, or a mixture, is precisely designed to achieve optimal acoustics for the instrument.

Brass instruments rely upon the natural harmonic series of the tube. The skill of a brass musician resides largely in her ability to adjust her **embouchure** (the position of the mouth and lips) and air pressure to pick out the desired harmonic accurately and reliably. This is particularly difficult for French horn players where the harmonic series becomes very narrow at the high end of its range. Perhaps you've heard a brass player miss a note and instead play an incorrect harmonic. So-called "natural" brass instruments of several centuries ago had no valves, much like the bugle of today, but still produced fine music thanks to the skill of the musician.

The Shape of the Pipe (Tube) Determines Its Acoustics

Finally, it's worthwhile here to explain some considerations for the design of woodwinds and brasses that are based on the physics of standing waves. It turns out that the relative intensity of harmonics produced in a tube (physicists usually call it a pipe because much of this discovery work and mathematical description was done with organ pipes) depends on whether one or both ends open to the atmosphere. It also depends on the shape of the pipe, especially whether it is a near uniform cylinder or has been designed with a conical bore. These two parameters have a large effect on the sound quality of the instrument and its range of pitches.

The physics for all of this is well understood and has to do with the position of **nodes** and **antinodes** in the vibrating column of air within the pipe. A node is a position in the standing wave where the molecules appear not to move with the compression and rarefaction of the air molecules. An antinode is a position where the molecules are compressing and de-compressing cyclically. The open end of the pipe can only have an antinode, whereas the closed end of a pipe can only have a node. This in turn determines the mix of harmonics that are permitted in the pipe.

These physical principles have major implications for the musician. For example, the flute and clarinet are approximately the same length and both have approximately cylindrical bores. Yet, the clarinet is able to play a fundamental pitch a sixth lower than the flute and has a very different mix of harmonics. The reason for this is that the flute is an open pipe (open at both ends), whereas the clarinet is a closed pipe (closed at one end by the vibrating reed and the airway of the performer). The physics of vibrating air columns shows that a closed cylindrical pipe is capable of generating only the odd harmonics, 1, 3, 5, 7 etc., whereas the open cylindrical pipe generates the entire harmonic series. A waveform produced electronically with only odd harmonics mimics the clarinet sound.[8]

The bore of wind instruments is complex because few instruments are made with a perfectly conical or perfectly cylindrical bore. For example, a well-designed professional clarinet has areas of narrowing in the bore as well as a flare at the bell end of the instrument. These deviations are designed by the manufacturer to improve overall intonation and response and in part account for the difference in performance of clarinets made by different manufacturers. The difference that the overall shape of the bore produces is illustrated by the difference between the sound of the similarly sized clarinet and soprano saxophone, with its conical bore. Both instruments use a single reed and are therefore closed pipes. The soprano saxophone, however, produces the entire harmonic series, unlike the clarinet's predominantly odd harmonics. Moreover, the clarinet produces its lowest pitch three whole steps below that of the soprano saxophone.

The Percussion

Now, we're ready to move on to a representative of the last group of instruments in the orchestra. We find those musicians sitting or standing in the far back row, as if relegated to the "cheap seats." The timpanist has been waiting for us graciously, more than ready to explain the wonders and diversity of percussion instruments.

He begins by explaining that this broad category of musical instruments has one thing in common: the performer strikes them (or sometimes scrapes or shakes them). Literally. This means that the ultimate driver is the muscle of the performer's arm, leg, hand, or fingers using some kind of striker (which sometimes can actually be the hand or fingers of the performer) impacting an object, which then vibrates as a result of the energy transferred to it—hence, the name **percussion**, derived from the Latin, *percussio*, meaning "the act of striking."

This is a very broad family of highly diverse instruments. They fall into two primary categories: pitched and non-pitched. Pitched percussion instruments produce definite frequencies of sound that belong to the musical scale. Non-pitched instruments do not. Percussion instruments can be further categorized into the following groups:

Idiophones: produce sound through vibration of the entire body of the instrument.

Membranophones: produce sound through the vibration of a membrane.

Chordophones: produce sound through the striking (as opposed to plucking or bowing) of a string.

Aerophones: specialized wind instruments generally played only by percussionists.

Below are examples of each type. The variety alone is impressive. In recent years, composers have begun to experiment broadly with the use of various percussion techniques, asking musicians to strike objects that might not be immediately recognizable as a musical instrument (such as a brake drum from an automobile or a jawbone from an animal) but such objects become a musical instrument when in the hands of a percussionist, however bizarre it may seem.

	Idiophone	Membranophone	Chordophone	Aerophone
Pitched	Chimes	Timpani (Kettledrums)	Piano	Slide whistle
	Handbells	Roto-Toms		
	Marimba			
	Celesta			
	Tuned Triangle			
	Xylophone			
	Glockenspiel			
	Crotales			
	Flexatone			
	Gongs			
	Musical Saw			
	Steel Drums			
	Vibraphone			
Non-pitched	Anvil	Bass Drum		
	Bell Tree	Snare Drum		
	Tam-Tam	Bongo Drum		
	Bata Drum	Congo Drums		
	Chocallo	Tambourine		
	Cowbell	Tenor Drums		
	Cymbals	Tom-Toms		
	Reco-Reco			
	Sleigh Bells			
	Auto Brake Drum			
	Thunder Sheet			
	Triangle			
	Cabaza			
	Castanets			
	Claves			
	Guiro			
	Jawbone (donkey, horse, or mule cattle)			
	Maracas			
	Rainstick			
	Ratchet			
	Sand Blocks			
	Slapstick			
	Temple Blocks			
	Woodblock			

Percussion instruments play a major role in contemporary music. They were also arguably the first types of instruments ever invented by musical humans and continue to be essential for music-making in cultures throughout the world. The combination of these pitched and non-pitched percussion instruments provides many opportunities for tone color, emphasis, rhythmic effects, sound variety, and even melody.

Before leaving our timpanist, he reminds us that the physics of sound still applies to all of these instruments, whether pitched or non-pitched. In each case, some kind of striker is used to set the instrument into vibration. Most produce quite complex waveforms as a result of their internal resonance. The quality of the sound is in part dependent on the striker, whether it is hard or soft, and how it's shaped. So, like the bow of string instruments, the device that transfers energy from human muscle to the vibrator is of critical importance here as well. Our timpanist also mentions that many percussion instruments are made of botanical materials, some of which are quite rare. He shows us the marimba as a beautiful example, where the vibrating bars, with their wonderful ethereal timbre, are made from rosewood (*Dalbergia stevensonii)*.

Marimba

Rosewood tree

Because percussionists are so involved in unusual means of generating musical sounds, he also reminds us not to forget various electronic instruments that are rapidly becoming more widely employed, even in classical music, but certainly by other musicians. These use an electronic oscillator, with its harmonic mix tuned precisely. While sound production is most often controlled using a standard keyboard, other means can be used as well. In this case, human muscle regulates the flow of electrons as opposed to providing the energy necessary to start the vibration. Similarly, the modern pipe organ employs electric pumps to push air through pipes, with a keyboard that enables the musician to choose the pipes that are activated. Organ flue pipes use an air reed, just as do piccolos and flutes. Most organs possess reed pipes as well, which are characterized by a vibrating strip of brass. To be complete here, metal reeds driven by air also constitute the vibrator for accordions and harmonicas.

In summary, then, musical instruments have much in common with each other as explained by science. Both the physical sciences and mathematics have contributed to their design, and certainly to our understanding of the ways that musical instruments work. Biology continues to contribute to the preservation of our music-making abilities because so many critical materials for our instruments are derived from plants, ranging from reed cane for woodwinds, to the entire bodies of string and woodwind instruments, to bows, to the vibrating portions of percussion instruments, and even to the wooden drumsticks that percussionists rely upon. Plant-derived materials are found everywhere in the modern orchestra. Chemists and engineers come into play as they try to find more durable substitutes for these plant materials, such as polypropylene instead of cane reeds and carbon fiber bodies for violins, and as they develop novel metallurgy to make better keys for woodwinds and better alloys for brass instruments.

Now that we understand this level of interaction between music-making and the physical, chemical, and biological sciences, engineering, and mathematics, let's proceed to the science of musical perception in us, as human beings. After all, it is we who create music and we who appreciate it, thanks to our amazing auditory system and, of course, our brains.

CHAPTER 8

The Ear

"Great music is in the ear of the beholder."
Noel Gallagher
Singer, songwriter, guitarist

"When music fails to agree to the ear, to soothe the ear and the heart and the senses, then it has missed the point."
Maria Callas
Soprano

We humans hear music by using reptilian jaw bones.

Yes, you've read the sentence correctly.

We humans can hear music only with bones that were once part of the reptilian jaw. More precisely, the bones were once part of the joint of the jaw of our reptilian ancestors but they were re-purposed through evolution to become components of our mammalian ears, thereby improving the fidelity, range, and the sensitivity of our auditory capabilities.[1]

Nature is an effective tinkerer. Throughout the course of the evolution of life on earth, she has routinely re-purposed many things: genes[2] acquired new functions, transcription factors (which regulate the activities of many other genes) mutated and altered the timing of the expression of whole batteries of genes, genes were duplicated and then one of those duplicated genes mutated to function in an entirely new way, proteins were retooled for a new purpose, bones were moved from one place to another, fins were transformed into limbs, limbs were transformed into wings, senses changed, and on and on. All of this is accomplished through random mutation (changes in genes and their regulators), followed by natural selection (the selection of advantageous changes that were then maintained in the population). While this tinkering sometimes required millions of years, like all tinkerers, Nature does make things work in the end—though they may not work efficiently or without some sort of compromise, or not necessarily in a way that we'd expect if we'd had a chance to design it ourselves. Indeed, we certainly find no evidence whatever for any kind of design in Nature's tinkering. Based on the abundant verifiable evidence, Nature is simply not a good design engineer.[3]

Darwin's ideas–now supported consistently by the findings of contemporary genetics, molecular biology, biochemistry, embryology, paleontology, comparative anatomy, comparative physiology, comparative biochemistry, genomics, plant biology, cell biology, and developmental biology—elegantly offer falsifiable and predictive explanation for the gradual changes in forms of life on earth. The origin of the highly sensitive ear of terrestrial mammals, which provides the best audition found on earth, is an especially remarkable and detailed narrative of evolutionary biology that is very well supported by abundant verifiable evidence first from embryology, secondly from the fossil record, thirdly and very recently from detailed anatomical studies on the development of at least one living species of marsupial, and finally through the immense power of molecular biology. This is one of the most stunningly well-documented examples of re-purposing in all of evolutionary biology.

The fossil record contains clear transitional forms that tell a compelling story of these bones gradually becoming smaller, losing functional importance in the jaws of a certain group of reptiles, and subsequently migrating a very short distance from the point where the jaw contacted the skull into the area of the middle ear. The bones became part of an evolving mammalian auditory system that was much more sensitive and far better adapted for the perception of higher frequencies of sound than the reptilian form, with its one bone, instead of three. This change occurred in a branch of ancient reptiles, as they became mammals, beginning at least 200 million years ago.[4]

The evolutionary journey is dramatically recapitulated in the very early development of the ears of young present-day gray short-tailed opossums, where we can observe bones from their jaws becoming bones in their ears over a period of a few weeks immediately after they are born. In essence, we actually see part of the evolution of the mammalian ear under laboratory conditions in a living animal, rather than relying solely on the fossils of extinct forms. And we even know something about the genes that are likely to be involved.[5]

So, what are these auditory bones, where are they found, and how do they function to make the human ear so exquisitely sensitive to sound waves?

Before we can inquire into the answers to those questions, let's learn some basic biology together. We have five principal senses: sight, hearing, taste, touch, and smell. In addition, we can perceive internal inputs as well, such as the position of our bodies in space, our balance, our state of nutrition, and felt vibration. The five principal senses have a common mechanism. Each one receives a physical stimulus from the environment and **transduces** it into the electrochemical signaling system used by neurons. Our eyes respond to the electromagnetic waves of light. Our ears respond to compression waves of sound. Our tongues respond to the physical presence of a wide variety of chemicals. Our skin and other touch sensitive surfaces respond to mechanical or temperature stimulation. And the olfactory areas of our noses respond to thousands of different airborne molecules.

The key to understanding our sense of hearing is this concept of **transduction.** The function of our ears is to gather the energy of physical sound waves and to change that information—to transduce the energy of these physical compressions and rarefactions of air molecules—first into mechanical energy and then ultimately into the energy of electrochemical signals that can be processed in our brains. Our ears do this remarkably well and with astonishing fidelity and great sensitivity.

So, how do our ears accomplish this wonderful task for us?

Let's look at the internal anatomy of the human ear[6]:

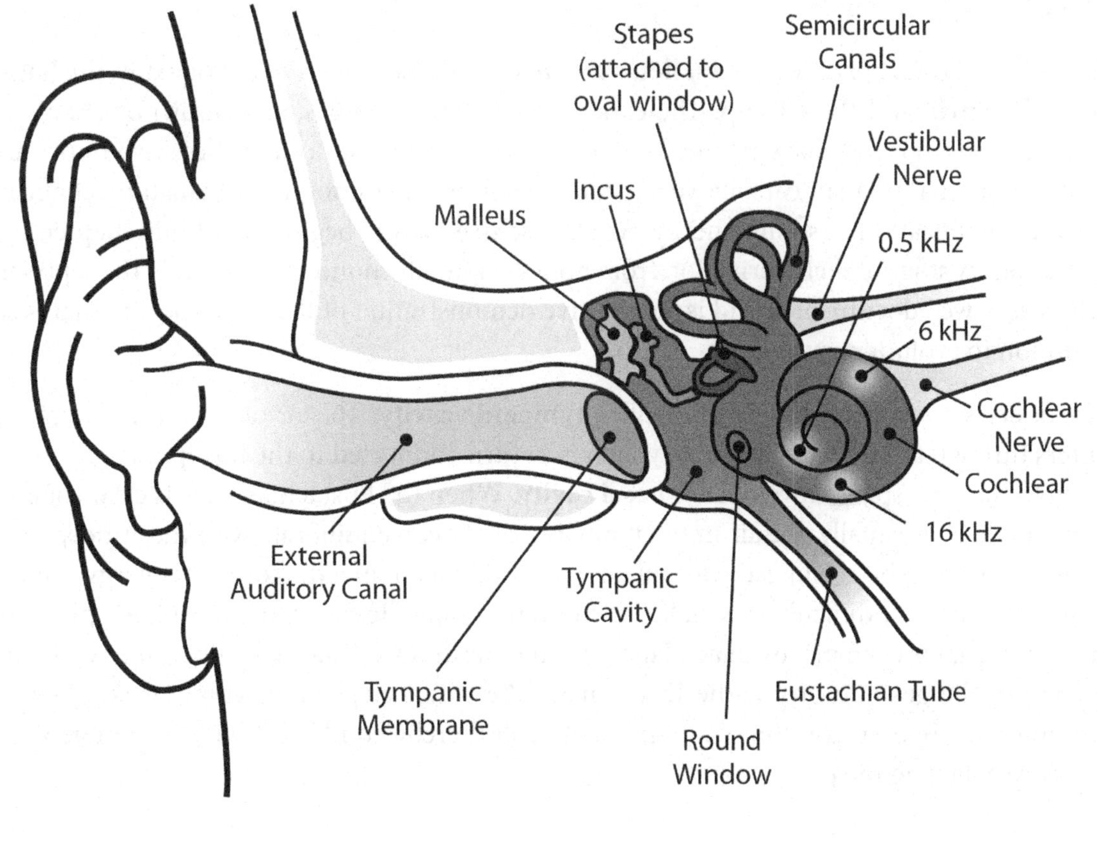

The human ear is divided into three areas: **the outer, the middle, and the inner ear**. The outer ear includes the **auricle (pinna),** which is the cartilaginous structure that we identify visually as "the ear." It gathers sound waves and focuses them into the **external auditory canal.** While many animals are able to move their pinnae (plural of pinna) to assist in determining the origin of a sound directionally, humans have lost that ability during the course of our evolution. The external auditory canal possesses glands that secrete earwax and also has small hairs, both of which protect the middle and inner ear.

The middle ear includes the **tympanic membrane,** known colloquially and descriptively as the eardrum. This membrane has three layers of tissue: the outermost is a thin layer of skin, the middle layer is fibrous tissue, and the inner layer is mucosal tissue, similar to the lining of the mouth. The tympanic membrane is subject to perforation as a result of infection or trauma such as inserting a cotton-tipped swab too far into the ear canal. Don't do that! Like the membrane of a musical drum, the tympanic membrane vibrates sympathetically to the sounds coming through the ear canal. These vibrations are then transferred mechanically to the **ossicles**, three small bones that amplify the energy received from the tympanic membrane by 15- to 20-fold.

These three bones are called the **malleus**, the **incus**, and the **stapes**, or colloquially, the **hammer**, the **anvil**, and the **stirrup.** The ears of amphibians, reptiles, and birds possess a stapes-like bone, but they have neither the malleus nor the incus. Both of these two bones are found in all mammalian ears, however, and constitute a common mammalian mechanism for the amplification of sound energy. As we discussed at the beginning of this chapter, the evolutionary story of their derivation from bones originally found in the joint of the reptilian jaw is very well documented and is a definitive demonstration of the re-purposing that is so common in evolutionary biology.

The ossicles are found within the air-filled **tympanic cavity**. The air of the tympanic cavity enters through the **Eustachian tube**, which is in turn connected to the **nasopharynx**, and is open to the atmosphere through the **nasal cavity**. When the Eustachian tube is clear (open), as it should be normally, the air in the tympanic cavity can equilibrate with the atmosphere. However, if the tube is plugged with mucus as a result of an infection or nasal allergy, equilibration can be very difficult, especially in children. A rapid decrease in atmospheric pressure encountered, for example, in an airplane, can then result in substantial pain as the tympanic membrane bulges outward. If the Eustachian tube is not clogged, yawning or swallowing will normally open it and the tympanic cavity will thereby equilibrate with the ambient air pressure, relieving the pain.

Ultimately, the vibration of the tympanic membrane is transferred through the motion of the ossicles to the **cochlea**, a lymph fluid-filled structure in the inner ear. The energy of sound waves, gathered and directed by the auricle, therefore flows from the air to a sympathetically vibrating membrane, then to a lever system that both amplifies and focuses the energy by reducing the relatively large surface of the tympanic membrane to the much smaller surface area of the stapes as it contacts the **oval window** of the cochlea. The physics of sound transmission makes this necessary because so much energy would be lost with a direct transfer of sound energy from air to fluid. In essence, this system matches the **impedance** (think of this as the resistance to vibration) of sound energy in these two media. The small surface area of the stapes concentrates the mechanical sound energy, thereby allowing the perception of sounds of much higher frequency.

The cochlea is shaped like a spiral and is quite an amazing outcome of natural selection. As the stapes moves against the oval window of the cochlea, the cochlear fluid begins to move. Within the cochlea are very small hairs that move sympathetically to the motion of the fluid. The flexible **round window** permits the fluid within the cochlea to move as vibration, in turn ensuring that the hairs will also move. Without the round window, the fluid would simply change the internal pressure of the cochlea.

The cochlear hairs are part of the **organ of Corti,** which some refer to as a microphone because it transduces the mechanical energy of vibration into the electrochemical energy employed by the brain. Hair cells are located throughout the cochlea and the perceived pitch of sound is apparently determined by which ones are stimulated. Hair cells closest to the stapes respond to the highest frequencies, as shown in the diagram (16 kHz), whereas those further away respond to mid-range frequencies (6 kHz), and those yet further away from the stapes respond to yet lower frequencies (0.5 kHz). The neurons from the cochlea are bundled into the **cochlear nerve**, which then transmits the signals to the brain. We can draw a schematic diagram that summarizes these points:

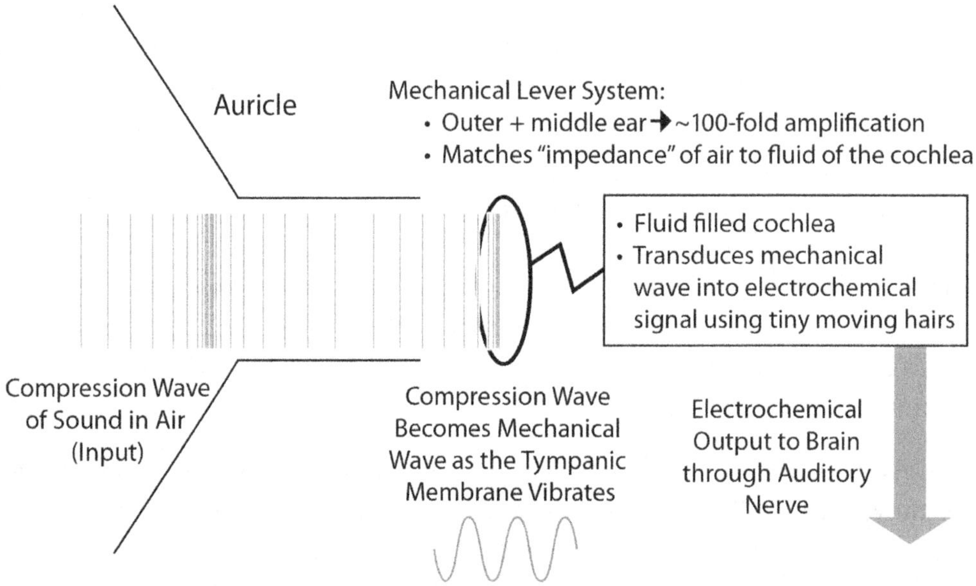

The important concept here is simple: the compression / rarefaction sound wave, gathered by the auricle into the external auditory canal, is first converted into a mechanical wave of vibration as it strikes the tympanic membrane. That energy is in turn transmitted through the motion of the ossicles to the cochlea. This entire process results in as much as a 100-fold amplification of the energy of the original sound wave to the point where it activates the hairs within the cochlea. This level of amplification is remarkable and makes it possible for us to perceive very soft sounds over a wide range of frequencies.

One structure of the inner ear not yet mentioned is the **semicircular canals**, involved in maintaining balance. In rather simplified terms, a change in the position of our head results in a change in the pressure of fluid in the semicircular canals as it presses against different receptors in the organ as a result of gravity. The pressure of fluid on different portions of the semicircular canals is transduced into an electrochemical signal, transmitted to the brain, which then uses that information to ensure balance. If something goes wrong with this system, we often experience vertigo or dizziness and can become physically ill.

Hurrah for Evolution!

If we think even a bit about the evolution of mammals, the requirement for highly sensitive hearing becomes quickly apparent. During the era of the great reptilian dinosaurs, most mammals remained small in size. As relatively small creatures, living in a world dominated by enormous dinosaurs everywhere around them, acute hearing would have provided a highly significant survival advantage, both in terms of detecting the presence of predators, as well as in finding food. Possessing three ossicles instead of only one enabled the auditory mechanism to be far more sensitive and provided a wider range of detectable frequencies, in part because of the amplification of sound energy. With the demise of the dinosaurs, the small mammals became larger, occupying then vacant ecological niches. As mammals, we humans are the direct and very fortunate beneficiaries of natural selection as it enabled highly sensitive auditory perception and processing in that ancient branch of reptiles as they evolved into the earliest mammals. About 25 million years ago, these evolved into the first primates, and about 7 million years ago, a branch of primates evolved into the last common ancestor for chimpanzees, bonobos, and hominids.

In point of fact, the survival advantages conferred by excellent audition in our mammalian ancestors of 200 million years ago now gives **us** a necessary component for our capacity to appreciate music. Hurrah for evolution!

A superior auditory system must have at least two properties. First, it must be highly sensitive to low level sound over a wide range of frequencies. Second, it must transmit, that is, present those sounds to the brain faithfully. Like a well-designed microphone, the auditory system must represent the sound with extremely high fidelity, accurately transmitting information about pitch, loudness, and timbre electrically.

The Physics of Measuring Sound Intensity (Loudness)

Let's evaluate these characteristics with regard to the human ear. First, regarding auditory sensitivity, that is, the perception of softness and loudness of the sound, we'll have to learn how physicists measure and express the loudness of sound and then relate that to the way musicians think about it.

First, physicists have traditionally used a unit called the **decibel** or dB. It's defined as follows:

$$\boldsymbol{\beta = 10 \log_{10} (I / 10^{-12} \text{ watts} / \text{m}^2)}$$

where β is decibels and I is the intensity of the sound wave measured in watts per square meter.

For those of us who are math-challenged, the details of the equation are less important than the concept. First off, the "10" is in the right side of the equation because we are calculating **deci** bels rather than merely bels. The intensity (I) of sound is the energy of the wave per unit of area. Physicists commonly use the watt as a measure of energy and the square meter as a measure of area.

Now, for the important thing about this equation: the quantity 10^{-12} watts/ m^2 is the smallest amount of sound intensity perceptible by a normal human ear. It is a tiny amount of energy. For those unfamiliar with exponents, it would be 0.000000000001 watts / m^2. A painfully dangerous level of sound intensity for us would be approximately 1 to 10 watts / m^2. The human ear therefore perceives loudness over many **orders of magnitude**[7], meaning a vast difference in intensity between softest and loudest. For that reason, this equation defines decibel or dB on a **logarithmic scale**. On this scale, that painfully loud sound at 1 to 10 watts / m^2 would be defined as 120 to 130 dB. Of course, as defined by the equation, the softest sound perceptible by a human ear would be 0 dB.

The dB is a very useful measure of loudness in many contexts. But it has a major flaw. The sensitivity to loudness of the human ear is not merely dependent on the intensity (the energy per unit area) of the wave but also on the **frequency** of the wave. In other words, our ears are more sensitive to sound at certain pitches than at others.

Scientists have developed another unit of loudness measurement called the **phon**. The phon relies on the **perception** of loudness and is defined as the number of dB measured using a tone with frequency of 1,000 Hz or 1 kHz. The normal human ear is very sensitive to frequencies in that range. Thus, the phon normalizes the dB value to a particular frequency. While the phon is an improvement over the dB in some ways as a quantitative estimate of perceived loudness, it still isn't perfect.

So, how do musicians describe their perception of loudness and how do composers indicate how loudly music should be performed? Do you remember all the Italian terms for tempo in Chapter 3? Well, musicians have Italian terms for loudness, which we call the **dynamic** of the sound:

ppp = *piano pianissimo* = extremely soft

pp = *pianissimo* = very soft

p = *piano* = soft

mp = *mezzo piano* = moderately soft

mf = *mezzo forte* = moderately loud

f = *forte* = loud

ff = *fortissimo* = very loud

fff = *forte fortissimo* = extremely loud

A point of humor: those of us musicians who are not brass players would add that brass players (especially trumpet players) commonly employ yet another, rather unpleasant, dynamic level, namely, *blastissimo*. Indeed, they seem to compete with each other to achieve ever higher levels of *blastissimo.* It appears to be an "ego thing."

Composers use additional symbols and words to describe a gradual change in dynamic. For example, *crescendo* is used to ask the performer to get louder gradually and *decrescendo* or *diminuendo* is used to indicate a gradually softer sound. *Subito* is an Italian word used to ask the performer to become louder or softer suddenly. All of these dynamic markings are subject to the artistic interpretation of the performer, always accounting for the size of the ensemble, the acoustics of the hall, the type of instruments, and many other factors.

How do these words relate to a more precise, quantitative measure of loudness? To answer that question, let's consider the **sone scale**, which was devised to connect the phon scale with the perceived loudness of the symphony orchestra, approximately in the range of 40 to 100 phons. It was also designed to correlate with our human perception of loudness using a linear as opposed to the logarithmic dB scale. In other words, a doubling in sones means a doubling in perceived loudness. Given that assumption, here is a chart comparing these measurements of loudness. Note that *mp* and *mf* are not distinguished in this chart.

Orchestral Dynamic	Phons	Sones
ppp	40	1
pp	50	2
p	60	4
mp	70	8
mf	70	8
f	80	16
ff	90	32
fff	100	64

As you've probably figured out by now, no loudness measurement system is perfect and this is the reason that we have so many of them. Yet another scale relies upon **sound pressure level**, as opposed to intensity (energy) but is still expressed in dB. The important realization for us here is that our ears are amazingly sensitive to very soft sounds and yet they have a capacity for the perception of much louder sound, regardless of the metric used to quantitate it.

Earlier we mentioned that the human ear is highly sensitive to frequencies of about 1 to 4 kHz. Actually, a normal, young person can hear frequencies between about 20 Hz and 20,000 Hz and can distinguish frequency differences of 1 Hz. As we grow older, we nearly always lose high frequency perception, so that at 70 years of age, the highest frequency heard, even at loud volume, is about 10,000 Hz.

How do we lose high frequency hearing first? The widely accepted explanation is that the energy of vibration coming from the stapes into the cochlea encounters the hairs that respond to high frequencies first. Therefore, they are subject to the full brunt of sound energy, and may well be the first to be damaged by loud noises. We do know that high levels of sound energy damage our hearing, something discovered by many rock stars. Today, orchestral musicians are much more aware of this danger than they've been in the past, and sound protection is mandated to preserve their hearing.

Finally, regarding the fidelity of our auditory system, we know that we distinguish pitch, loudness, and timbre very well because of results from hearing tests. Apparently, although we can never know for certain, all human beings hear a saxophone as different from a clarinet in the same way. If that's the case, the ear is indeed a high-fidelity transducer. On the other hand, we do have evidence that indicates that the size, density, and shape of person's skull influences its tendency to resonate with various pitches, in turn influencing the way that the ear perceives a particular sound. Of course, the structure of the ear itself is likely to differ a bit from person to person, and that could influence perception as well. Much research is still required to document human variations in auditory fidelity.

In this chapter, we've come a long way from reptilian jaw bones to an understanding of our amazing auditory capabilities. But the best ears in the world mean nothing without the processing of those electrochemical signals in the brain. Indeed, it's well worth repeating many times in this book: music does not exist as sound waves in air. Music certainly doesn't exist as printed symbols on paper.

Music exists only in the complex neuronal activities of our own brains.

That's where music lives. That's where it has its form and its reality.

So, let's investigate the human brain!

CHAPTER 9

Our Musical Brain

"The human brain, then, is the most complicated organization of matter that we know."

Isaac Asimov

Biochemist, Writer, Futurist

"The brain is a three-pound mass you can hold in the palm of your hand that can conceive of a Universe 100-billion light years across."

Marian Diamond

Neuroscientist

As I was driving home just at dusk in late summer after visiting a bassoonist friend's country home in Missouri—with the weather unsettled and the humid air hanging heavily—I was startled to see directly in front of me a single, isolated storm cloud, and nothing else in the sky. With the setting sun's light outlining this lone cloud, it seemed in my imagination to be shaped very much like a human brain, floating in midair, as if suspended. Inside the dark gray cloud, electrical sheet lightning[1] was constantly being generated and firing apparently at random, illuminating the cloud internally. Occasionally, lightning would escape and strike the earth, but most of the lightning remained within the cloud. The intense white light of the electrical discharges contrasted with the red glow from the waning sun to create an ethereally beautiful effect.

Suddenly, it occurred to me that this was indeed a visual metaphor for our own brains as we make music. I already knew from recent research with sophisticated imaging technologies that our entire brain "lights up" when we make music. "Lights up" is an apt term used by neuroscientists when they observe enhanced activities in specific areas of the brain. Insofar as neuroscience can ascertain, nothing puts more demand on the capabilities of our brains than making music.[2]

To get some idea for how this assertion makes sense, let's do another *Gedankenexperiment*. Remember that science begins with imaginative inquiry. So, imagine with me for a moment that you are a professional musician performing a concert with a highly accomplished symphony orchestra and a world-renowned conductor. After many thousands of hours of practice and instruction from some of the finest conservatory musicians in the world, you are now expert with your instrument, well prepared and ready to perform. The conductor lifts his baton to begin. What does your brain now have to do? Which senses will be involved? What must happen?

First, your visual sense is engaged. You see the conductor. You read the written music in front of you. Somehow, your brain must process those symbols on paper and convert that information instantaneously into the motion of your muscles. Your respiration, fine finger control, arm motion, body posture, tactile sensations, mouth muscles (if you play a wind instrument) and vocal muscles (if you are a singer) all must work synchronously and flawlessly.

At the same time, you are also watching the conductor to ensure that you are playing according to the tempo that he is setting. You are listening constantly and carefully to evaluate your own sound to ensure that it is exactly what you want to produce as indicated by the written symbols in front of you. Of course, you are also listening to the one hundred other musicians of the orchestra to ensure that you are playing in tune with them and that your internal pulse always matches the pulse of the orchestra. You must switch your attention constantly from instruments playing your particular musical line to those setting the underlying pulse to those playing a counter melody to those with whom you must be precisely in tune, back to the conductor, and on and on for long stretches of time.

You use all of that information perceived from your senses to make constant and innumerable "course corrections" in your playing within tiny fractions of a second during the entire concert—a concert that may continue for well over two hours. You are also continuously planning ahead, anticipating the next part of the music to be played, and processing any unexpected deviations. And, then, in addition to all of this executive and planning function, with the visual, auditory, tactile, kinesthetic sensing and feedback processing, you are using the core of your emotional sensibility to make certain that the sound that you produce is emotionally evocative and that your musical style, dynamics, rhythm, and phrasing all create the emotion that you want to create, as demanded by the composition and as indicated by the conductor.

That's a lot for one brain to do!

In fact, merely listening to music intently and attentively activates many of the same areas of the brain, though not all. Making music, especially at a high level of expertise, alters the **physical anatomical structure**[3] of the brain, whereas merely listening to music appears not to do so in the same way.

In this chapter, we'll inquire together about several aspects of the human brain, with special emphasis on its perception and processing of music. To achieve that goal, we must review a bit about brain anatomy, not in the sense of a textbook, but in the sense of understanding the brain's general functional anatomy in an accessible way. In other words, we'll discuss enough brain anatomy to understand the landmarks without aiming to prepare you for a medical board exam in neurology. [A sigh of relief from you is entirely warranted here.]

We'll also critically evaluate the new tools used today to image living human brains. These tools, together with observations of the effects of various brain lesions and deficits caused by stroke, trauma, or congenital anomaly have given scientists the information necessary to begin to unravel the mysteries of music perception and processing. Finally, we'll review the growing scientific evidence for the location of specific music processing functions and their wide distribution throughout our brains, uncannily reminiscent of the widely distributed sheet lightning throughout that lone storm cloud in the Missouri summer evening sky.

Complexity of the Human Brain[4,5]

The human brain is enormously complex and some people reasonably argue that our brain can never fully understand itself. Maybe that's so. But at this point in the history of science, we can't know either way. Neuroscience is making progress at an unprecedented rate and those who in the past have had the temerity to doubt or to discount either the power of science or the persistence of scientists in seeking to understand Nature have most often been very surprised—and wrong.

The brain's complexity is found at many levels but one useful distinction is between the microscopic (small) and the macroscopic (large). The microscopic is at the level of cells and the physiologic and biochemical mechanisms by which **neurons**, or **nerve cells,** communicate with each other. The brain is composed of approximately 100 billion neurons, each with as many as 50,000 intercellular connections and perhaps on the order of 1,000 trillion connections in the entire nervous system. Such a number is difficult to fathom. All of these connections *in toto* have now been dubbed **the connectome.**

In addition to neurons, we find an equal number of **glial cells**, sometimes called simply **glia** in the brain. While glial cells come in a wide variety of types, the easiest way to think of them is as "helper cells" that supply nutrients to neurons, protect them from disease, provide appropriate insulation, and maintain structural scaffolding. Additional functions for glia continue to be identified. We can't spend much time here discussing glia because of our limited space, but they are important, especially if you happen to be a neuron.

The neuron is defined as an electrically excitable cell. In point of fact, all living cells rely upon electrochemistry to remain alive, but neurons are specialized in a way that began to evolve perhaps 500 million years ago in the first animals. Only sponges and a few other such very simple animals today lack neurons. All neurons function by efficiently receiving, processing, and transmitting electrical and chemical signals. Many types exist, including those involved specifically with our senses, those that regulate movement, and those that provide interconnections between other neurons.

Here's a simplified sketch of two typical neurons and the connection between them. Each neuron has a **cell body** that houses the life-sustaining basic metabolic activities of the cell, including the **nucleus**. Typically, projecting from the cell body are many branching **dendrites** and one longer end-branching **axon**. Axons can be very long. The longest axon in the human body extends from the base of the spine to the big toe.

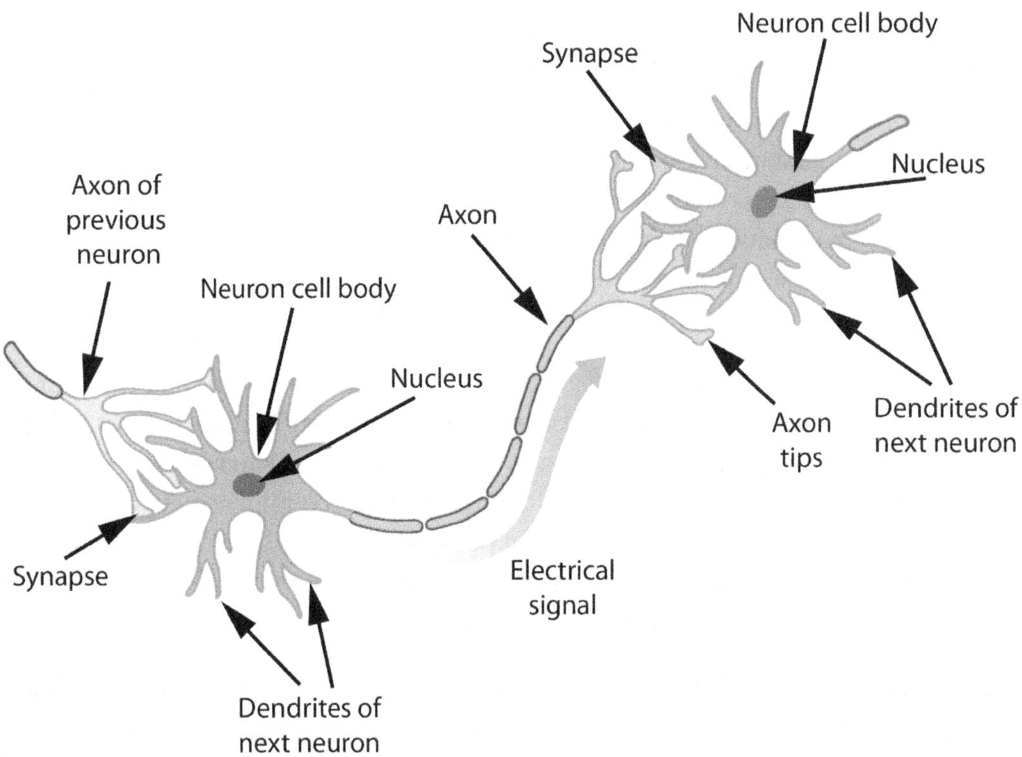

Signals are received by the neuron through its dendrites when they perceive presence of specific chemical mediators, called **neurotransmitters**[6], which may either promote information transfer or inhibit the activity of the neuron. These neurotransmitters are released by other neurons at the **synapse**, a very small space between the axon of one neuron and the dendrite of another. In the event that sufficient levels of appropriate neurotransmitters are present, the neuron undergoes **depolarization**, a change in electrical charge brought about by the transfer of ions across its membrane. This electrochemical depolarization travels down the axon as an electrical **action potential** and triggers the release of neurotransmitters at the branching ends of the axon, proximate to the dendrites of another neuron (that is, at another synapse). And the process of signal transmission thereby continues.

Cell bodies, together with their dendrites, as well as the ends of axons, appear to be gray in color when the brain is examined. Hence, this is called **gray matter** historically. On the other hand, axons, bundled together and "insulated" with myelin (covered with special myelin-maintaining glial cells), appear to be white in color. Not surprisingly, these are called **white matter**. The distinction is easily visible in the dissected brain macroscopically (just by looking at structures visible with the naked eye) and especially upon microscopic inspection. The largest collection of myelinated axons in the brain is the **corpus callosum**, which connects the two hemispheres of the brain (the left and right.)

One key concept here is that if a signal between two neurons occurs repeatedly, over and over again, that synaptic signal becomes **potentiated**. In other words, the neuron fires more easily and probably more often. A better way to describe this may be that the signal causes the firing of its neighboring neuron more readily because neuronal firing is an all-or-none process. Either the firing occurs, or it does not. There is no "half" or "partial" firing. Neuroscientists think that the basis of memory is in the consolidation of these potentiated synapses, which is mediated in turn by permanent changes in the biochemical reactions within neurons, including the activities of specific genes.

Now this is all a bit simplified and generalized. There are many exceptions. Some neurons lack an axon. Some are connected to other neurons electrically without a neurotransmitter intermediary. Some neurons exist merely to connect one group of neurons with another. But the general pattern as outlined remains useful for our discussion here. Summarizing, it's important to remember that neuronal transmission is fundamentally an **electrical / chemical** phenomenon, generated by chemical changes in the cell, and that this electrical phenomenon can be **potentiated**.

Our Musical Brains Run on Electricity

The energy requirement for the brain is large relative to the rest of the body. Though the brain weighs only 3 pounds (less than 2% of the weight of a 175-pound person), it utilizes roughly 20% of the total calories that we consume. Its metabolic activity per unit weight is much higher than skeletal muscle and seems to increase with greater cognitive demand. (Brain over brawn!) In rough numbers, the brain requires 400 calories per day, the equivalent of continuously generating about 20 watts of power. Although making bioelectricity requires a lot of energy, our brains are far, far more efficient in terms of energy consumption per unit of computational power than even our best supercomputers. The human brain is definitely a remarkable product of evolution by natural selection.

Modules and Networks

We can well imagine that individual neurons acquire a specific function. Groups of neurons could become **modules,** all working together to achieve a particular task. Modules themselves would then acquire a higher-level specialized function, providing a mechanism for a division of labor, and increased efficiency within the entire brain.

A recent study[7] on the architecture of the human brain makes exactly this point. Using contemporary imaging technology, which we will discuss in much more detail momentarily, the authors of this study conclude that modules remain autonomous, though connected, with each module executing a cognitive function. If this model is correct, the overall larger anatomical architecture of the human brain becomes even more understandable because various modules could be housed in specific anatomical (geographical) areas of the brain.

The next level of organizational complexity of our musical brain is often thought of as the formation of **neural networks**. The exact details of this level of neuronal organization are subject to considerable conjecture. It appears that **hub nodes** act as a point of connection for multiple modules. Networks thereby connect one neuronal function with another—or one module with another—through **circuits**.

In summary, then, we could think of a hierarchy of organization:

1. The individual neuron;
2. Modules of neurons forming a "work group;"
3. Each module located in a particular part of the physical structure of the brain;
4. Networks of modules circuited dynamically through nodes to achieve specific tasks.

One admittedly imperfect analogy is to think of the brain's networks as being somewhat like the internet. You could consider your own personal laptop computer as a single neuron. It can connect with an untold number of other neurons (other computers owned by other individuals) by sending bundled information encoded in 0 / 1 binary code to computers termed "servers." They are analogous in some ways to the modules in your brain. The servers may possess encoded information of value to you, conduct computations, or by acting as a hub may connect you with yet other servers, and then with even more servers across the globe. The usefulness of this relies upon sufficient "bandwidth," that is, the ability to transmit sufficient amounts of information rapidly. It is like that with your brain. The connections between modular structures within your brain require sufficiently large capacities and sufficiently fast speeds of transmission.

As the Brain Makes Music

Now, let's combine an examination of the anatomy of the brain with what neuroscience has learned about the involvement of various parts of the brain in making music. We'll use only two views (drawing perspectives) of the brain for the sake of simplicity. The first is called a **medial view.** Imagine that we slice the brain in half, down the center. This allows us to view the internal anatomical structures of the brain. The second view will be a **side view** of the entire brain from its left side. While the brain can be dissected and viewed from many other angles, these two views will provide enough information for our inquiry here.

On a sketch of a medial view of the brain below, I've labeled four regions that provide us with a basic map, not at all detailed, but with signposts for future discussion. Brain structure at this level is hierarchical in at least two ways. First, it is hierarchical with regard to function. The brainstem region controls the basic functions critical for life and each succeeding step up the ladder toward the top of the skull becomes more sophisticated. Secondly, it is hierarchical from the standpoint of evolution. Whether we study the fossil record, embryology, or living animals of increasing complexity, it becomes clear that the brain stem region (or something like it) evolved first, followed by the midbrain, followed closely by the limbic region, and finally the cortical region. Note, however, that the brainstem, midbrain, and limbic regions are all ancient parts of vertebrate brain anatomy. The mammalian cortical region shows increasing complexity, reaching a peak in advanced primates, cetaceans (whales, dolphins), canids (dogs, wolves) and elephants.

Major Anatomical Divisions of the Human Brain (Medical View)

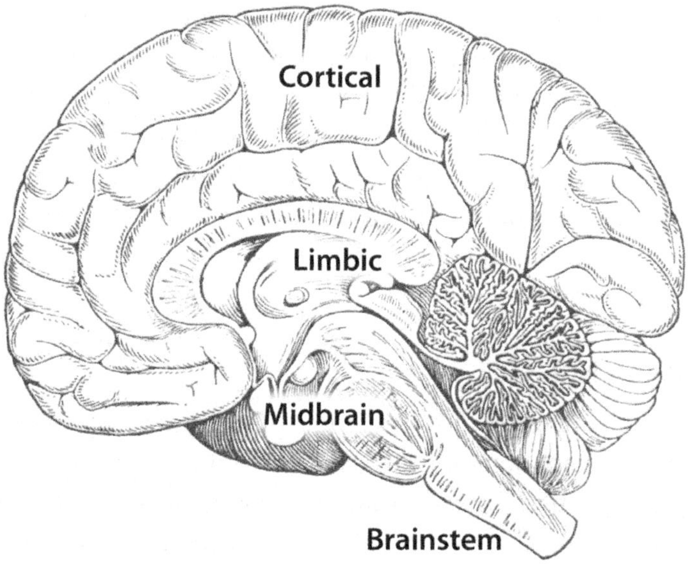

The basic functions of these four regions:

Cortical Region: *Planning, reasoning, working memory, abstract thought*

Limbic Region: *Long-term memory consolidation, emotion, instinct*

Midbrain Region: *Early processing of sensory and other signals*

Brainstem Region: *Autonomic functions essential for life* e.g. *respiration, heartbeat*

Reflected in this hierarchical organization, then, is a specialization of various areas of the brain. Remember that these areas are closely connected and must operate efficiently together.

Textbooks on the anatomy of the central nervous system provide many detailed and beautiful representations of brain structure. Online, you can view all sorts of details and if you are interested, I recommend them. Do remember, however, that various authors will often name parts of the brain in somewhat different ways, mostly for historical reasons. For our purposes here in this chapter, we'll provide simpler representations because that's all that we really need to describe the following major observations:

1. The performance of music requires the participation of many areas of the brain simultaneously;
2. The involvement of each specific area makes good rational sense from the standpoint of its ordinary, non-musical function; and
3. As we shall learn in Chapter 10, some areas of the brain change physically as a result of musical training and musical performance.

Imaging the Brain

Our first task is to understand the sources of verifiable evidence that provide confidence about the identification of the areas of the brain involved in making music. The earliest evidence could come only from crude X-rays or from post-mortem examination of the brains of patients who had either been injured, had a tumor, had suffered a stroke that resulted in the loss of particular functions, musical or otherwise, or had a congenital anomaly. While this may seem primitive, it continues even now to provide information about the localized functioning of many areas of our brains and is very useful, as far as it goes. The famous composer, Maurice Ravel, for example, suffered brain disease that made it impossible for him to transcribe the plethora of musical ideas that were still being generated by the remaining healthy parts of his brain, indicating that musical ideas and the ability to write them down reside in different areas of the brain.[8]

In the last few decades, an enormous advance has been made. We can now image **living** human brains with a variety of different tools, each with its own advantages and disadvantages. Moreover, we can image living brains as their owners engage in specific tasks or are at rest.

Here is a summary of those imaging technologies. Note that these technologies are divided into two basic categories, structural and functional. Structural imaging is static but it powerfully defines the structure of the brain, rapidly providing information to the clinician about injury, tumor, and abnormal anatomy. Functional imaging, the newer form of imaging, provides information about the real-time activities of all parts of the living brain of the patient and therefore provides the clinician with a broad, enormously useful set of diagnostic tools.

Tools for Brain Imaging

To Reveal Brain Structure Only:

Name	Employs to Image	Comment
Static X-ray Image	X-ray	Limited detail
Computerized Axial Tomography	X-ray	Good detail
—CAT or CT Scan		
—Images brain in slices (tomograms)		
Magnetic Resonance Imaging	Magnetic Field	Excellent detail
—MRI		No radioactivity
—Tomograms		No X-radiation
Cranial Ultrasound	Ultrasound waves	Useful in infants

To Reveal Brain Structure <u>and</u> Function:

Positron Emission Spectroscopy	Radioactive Sugar	Functional detail
—PET scan		Radioactivity
—Measures cell metabolism		Limited temporally
Functional MRI	Magnetic Field	Functional detail
—fMRI		Limited temporally
—Measures blood oxygen as marker		No radioactivity
		No X-radiation
Single Photon Emitted Computerized Tomography	Radioactive Isotope	Excellent detail
—SPECT		Radioactivity
—Measures blood flow as marker for activity		Temporally limited
Electroencephalography	Surface Electrodes	Excellent temporally
—EEG		Poor spatial detail
—Measures electrical activity		

Magnetoencephalography	Magnetometer	Excellent temporally
—MEG		Poor spatial detail
—Measures magnetic fields		
Diffuse Optical Imaging	Infrared Light	Less often used
—DOI (also Tomography, DOT)		Limited temporally
—Primarily measures blood flow as marker		Adjunct to MRI
Event-Related Optical Signaling	Infrared through	Less often used
—EROS	optical fibers	Only 1 cm deep
—Measures light scattering properties of neurons, not blood flow		
Functional Diffusion Tensor MRI	Magnetic Field	White matter studies
—fDTI		
—Measures distribution of water		

Obviously from this list, we now have many ways to image the brain, several of them potentially applicable while the brain is actually performing a task. The technologies most often used functionally are PET scans, fMRI, EEG, and MEG, though the others may be used more often in the future as clinicians and researchers become more familiar with their appropriate application. Also, new methods are constantly being invented and older ones improved.

Critical Assessment of Imaging Tools

As you may recall from Chapter 1, I committed to provide a skeptical assessment of all the science that we would discuss in this book. To that effect, I must state unequivocally that no imaging technology is perfect. Remember that the brain works in **milliseconds** and two of the technologies most widely used primarily because of their spatial resolution, PET and fMRI, detect brain events in **seconds or tens of seconds**, which is a 1000-fold difference. Moreover, PET and fMRI methods don't even measure the firing of neurons but rather measure surrogate markers for neuronal activity. [Aside: it's possible in some cases that we are not distinguishing the activity of glia from that of neurons.] All conclusions that we reach about the involvement of a particular area of the brain for a particular function must be made cautiously and provisionally. Moreover, each imaging technology has its own level of sensitivity, which can be adjusted. A given level of sensitivity may record a change that could later turn out to be an artifact, that is, a false reading that occurred as a result of a flaw in the measurement technique as opposed to an actual property of Nature. *Caveat emptor.* In other words, both the buyer and the reader here should beware. As we learned in Chapter 1, science is characterized by empiricism **and** skepticism.

The results can be visually stunning, however. For example, here are PET scans made when a patient is either listening to music or not. The color rendition is on the back cover of this book, where blue indicates less active, with green, yellow, and red indicating increasing activity of the area of the brain:

The brain at rest

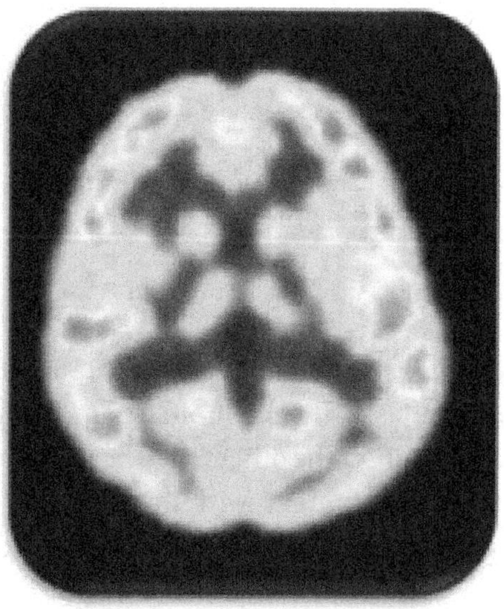

The brain's reaction to music

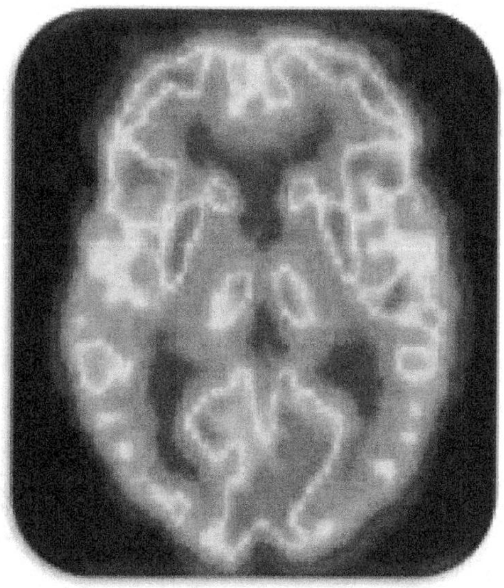

When such images are combined with data from animal studies (where electrical activity can be probed directly), studies on stroke patients, on trauma patients, extensive post-mortem examinations, and biochemical analysis, they have been very revealing. Also, as we'll discuss in more detail in Chapter 10, both EEG and MEG have much better temporal resolution than either PET or fMRI. Large amounts of data are constantly accumulating from clinical and non-clinical studies. The sheer volume of music studies in the neuroscientific literature indicates that neuroscientists consider the study of music perception and music-making especially useful in learning new things about the brain—or it could be that scientists are just interested in music—or perhaps both.

Regardless, the perception of music is really about the extraction of information content and meaning from sound. This is a very complex activity of the brain and the evidence clearly shows that this extraction involves many different anatomical structures and circuits. Let's study some of those anatomic structures.

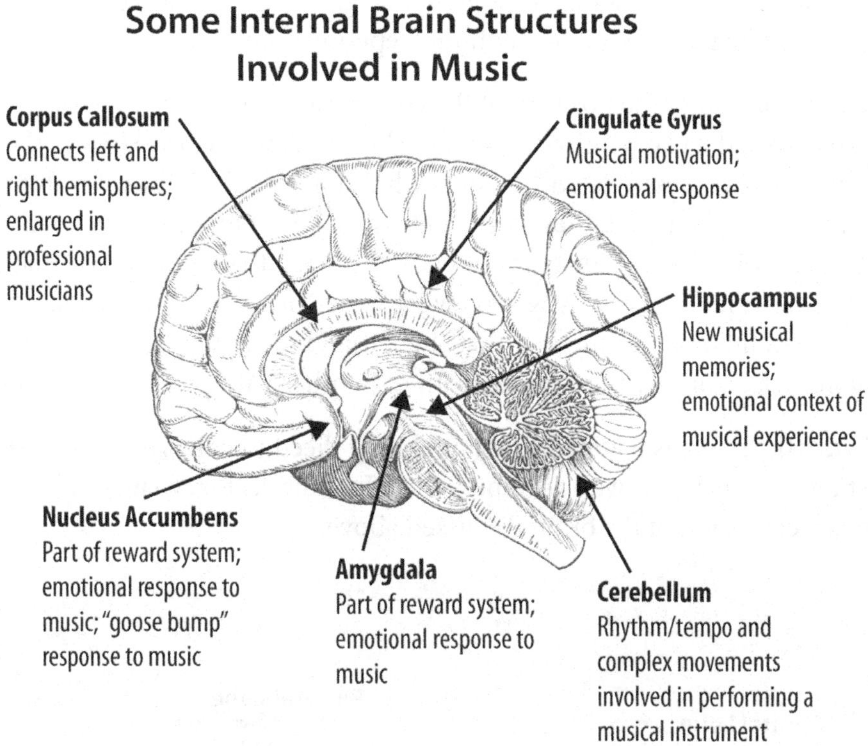

Diagram of medial view of the brain, with emphasis on the areas of the limbic system involved in the perception of music and in the assignment of meaning to music.

Imaging and other results have revealed increased activity with music in areas of the brain involved in emotional response and memory consolidation, namely the hippocampus, the amygdala, and the nucleus accumbens. These are part of the brain's "reward system" and since making music or listening to music triggers emotion, this is not unexpected. The exact names and precise locations of each of these regions can be a bit daunting to understand, in part because brain anatomy is so complex. The names are not nearly as important as the concept that the emotional and reward centers of the brain "light up" with music.[9]

The cerebellum is important in the control of movement and in the consolidation of patterns of movement, especially complex movements. From what we learned in Chapter 7 ("Making Music"), it is reasonable to postulate that all parts of the brain involved in the coordination of muscle activity, but especially the cerebellum, would be important for musicians, including vocalists. Our muscles are the drivers that start vibration. Even listening to music involves muscular activity. Think of dancing or foot tapping. To that point, the cerebellum "lights up" whenever a person hears any kind of rhythmic, repetitive sound.

Finally, the drawing shows the position of the corpus callosum. This is a structure that connects the two halves, the two hemispheres, of the brain. As we shall see below, the two hemispheres of the brain possess related but somewhat different functions, resulting in a division of labor between them. Yet, they must communicate. Failure to do so results in various well-defined pathologies and deficits in functional capabilities. It turns out that the corpus callosum of professional musicians is somewhat larger on average than that of non-musicians. This may well be due to the fact that the left and right hemispheres of the brains of performing musicians must communicate with each other with greater "bandwidth."

Next, we'll view the brain as it appears without being sliced in half. This view is from the left side of the brain and helps us with the anatomical landmarks for what is called the cerebral cortex, the cortical region of the brain discussed above.

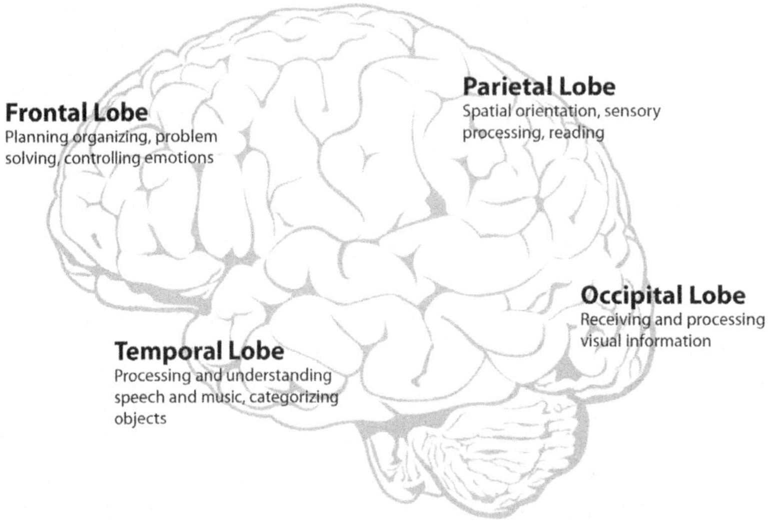

The cerebral cortex has four large parts, or lobes, each with its own particular set of functions:

- **The Frontal Lobe** is responsible for executive functions: planning, thinking, solving problems, foresight, anticipation, and controlling emotions.
- **The Parietal** (puh-RAHY-i-tl) **Lobe** is responsible for processing much somatosensory information and for reading.

- **The Temporal Lobe** helps to form new memories, categorizes objects and events, and processes both speech and music.
- **The Occipital** (ok-SIP-i-tl) **Lobe** receives and processes visual information.

Each lobe on the left side of the brain, which we are viewing here, has its counterpart in the right hemisphere. There is some specialization and division of labor between the hemispheres, in particular because the left hemisphere generally controls the right side of the body and the right hemisphere controls the left side. As we shall see, this specialization extends to music as well.

The scientific literature that describes the very complex anatomy and the functioning of various areas of the brain is enormous and growing. Many textbooks have been written on this topic. Of necessity, we will be somewhat cursory in our discussion here, but it is possible to set forth the highlights of what we know about the locations of music processing in our brains and our degree of confidence in our analysis. That said, we must always keep in mind that each human brain is different. Each one of us has somewhat different neuroanatomical architecture and circuitry and any conclusion ought to be considered a generalization.

Before we move forward, let's cover one very basic point, namely, the pathway of auditory signals from the cochlea to the temporal lobes of the brain where the auditory cortex is located, a critical area for processing all sounds, but especially for music and speech. The auditory nerves of both ears are first linked directly to specific auditory areas of the brainstem, where the initial processing of the signal takes place. The signals are then processed by specific areas of the midbrain and from there the auditory signals go to the thalamus, an area almost exactly in the center of the brain that acts as a relay station or hub for sensory signals (except olfaction). The thalamus directs the auditory signals to areas of both the left and right temporal lobes that are specifically involved in audition and in the interpretation of sounds. Thus, electrical signals from the cochlea are processed sequentially at all four major levels in the brain.

The Parts of The Brain Involved in Making Music[10]

We can choose to organize the massive amount of information about music processing in the brain now gathered by neuroscientists in either of two broad schemes. We could list each area of the brain that we know about and describe the evidence for its involvement in music perception and in music making. Or we could list the elements of music and then identify those areas of the brain that seem to be involved based upon evidence from any of the methods that we've described above. Let's use the elements of music to lead us to the regions involved in the brain.

- Perception of Pulse
  - The left temporal lobe auditory cortex processes rhythm.
  - The cerebellum assists in entrainment to the pulse.
  - The frontal lobe motor cortex engages with movement.
- Perception of Pitch
  - The left temporal lobe auditory cortex maps directly to the cochlea for the perception of pitch.
- Perception of Loudness
  - The temporal lobe auditory cortex responds to and perceives the loudness of sound and this correlates with psycho-acoustical measures of loudness as opposed to the physical sound pressure level.
- Perception of Timbre
  - The right temporal lobe auditory cortex processes and analyzes timbre.
- Processing of Melody
  - The right temporal lobe auditory cortex processes melody.
  - The hypothalamus is involved in determining whether music is pleasurable.
- Processing of Harmony
  - Both the right frontal cortex and part of the cingulate gyrus are involved in the appreciation of harmony.
- Emotional Reaction to Music
  - The amygdala is very important in the emotional connection to music and in connecting memories with specific music.
  - Other areas of the limbic region are also involved in the pleasure response.
- Performance of Music
  - Any area of the brain that processes pulse, loudness, timbre, melody, harmony, and emotional reaction to music.
  - The frontal lobe anticipates musical entrances, enables artistic interpretation, initiates motivation to practice; Broca's area, located in the left frontal lobe is involved in keeping time and singing lyrics.
  - The cingulate gyrus, part of the limbic region and lying just above the corpus callosum in the center of the brain, is involved with the stress associated with performance, deciding whether to practice, and attention to the task at hand.

- The motor cortex, at the rear of the frontal lobe, is responsible for many aspects of movement, including dexterity, so-called "muscle" memory, and rhythmic movements.
- The hippocampus enables the performance of an individual musical line, independent of the rest of the ensemble.
- The nucleus accumbens is involved in the emotional response to music and the "goose bump" response that many people have with music.
- The basal ganglia (part of the limbic region) is involved in holding the musical instrument, feeling the pulse, and memorizing music.
- The thalamus attends to the conductor and to the rest of the ensemble, and processes symbols of written music in order to execute motor movement.
- The parietal lobe (various parts) processes tempo, rhythm, dynamics, understanding groups of notes.
- The occipital lobe is necessary for reading music and for viewing conductor.

- Composition of Music[11]
    - All brain areas for all other musical functions.
    - Stronger synchronization between certain parts of the cingulate gyrus and a specific part of the frontal lobe and an area of the parietal lobe during the actual process of composition, whereas certain other connections are decreased in intensity.

One point here is particularly important. While the above outline does provide some insight into specific areas of the brain that neuroscientists have identified as involved in music perception, performance, and composition, it's critical to emphasize that no one brain area stands alone. All are **networked** together. We should actually think in terms of circuits of neuronal activity within the brain. One example is the limbic area where a circuit exists that leads to the perception of pleasure. But many other circuits exist between various areas of the brain, as identified with the imaging methods discussed above. Neuroscientists look for areas of the brain that light up simultaneously or sometimes sequentially with a given task. Such a finding indicates that those areas are circuited together and interact to achieve the task. Just as it takes a village to raise a child, it takes an entire brain, "wired" together, to make music.

Another point is also important. Musicians become expert only after many thousands of hours of arduous, intense practice under the guidance of good teachers. Expert classical music performance, in particular, cannot be self-taught. High level expertise can only be acquired from good teachers who not only teach the technical and artistic aspects of making music, but who also provide constant feedback and correction. While a certain level of inborn "talent" or musicality is essential, it's really the discipline and diligence of practice that create expertise.

How do neuroscientists think that this may happen? Evidence is accumulating that the perception of patterns and the coordination of movements are stored as "subroutines" within the brain. In other words, practice creates the potentiation of synapses in groups of neurons that are accessed when required to make music. Thus, a skilled musician recognizes a note pattern as a D-major scale almost unconsciously and the appropriate subroutines within his brain are accessed to perform that scale, even without conscious thought. This is a remarkable feature of neurophysiology that undoubtedly is a key enabler for making music.

Summary of Major Conclusions

In summary, neuroscience has accumulated verifiable evidence that making music involves many regions of the brain and the circuits that connect them. Here are certain general principles that may be helpful to consolidate our understanding:

1. The temporal lobes, and in particular the auditory cortex within those lobes process pitch, timbre, loudness, melody, and harmony;
2. Several areas of the brain, including the cerebellum, the parietal lobe, and the frontal lobe motor cortex all participate in entrainment to the pulse, probably because "feeling the pulse" is intimately tied to movement;
3. The frontal and parietal lobes of the cortex are involved in the executive functions of making music and in the control and initiation of necessary movements;
4. The occipital lobe is critical for reading music and for following the conductor and in small ensembles, for viewing the body movements of fellow performers to assist with maintenance of a common pulse;
5. The limbic region functions to provide the emotional context of music and the motivation for making music.

Music does not exist independent of the human brain, whereas sound (as sound) can easily be detected, demonstrated, measured, described, analyzed, and characterized with scientific devices alone. Music exists **only** in the firing of neurons in our brains as our brains interpret certain sounds as music.

In essence, our brains **create** what we call music from the sound waves striking our ears. This is a foundational realization and it poses many more questions. Given the human energy and effort that making music requires, what are the biological or other causes that drive us to invest so much effort? Does music change us? Anatomically? Emotionally? Cognitively?

Biologically? Psychologically? Physically? The central, underlying, and profound question here is simply:

Can we explain the reasons that humans invest so much effort in making music?

That is the subject of our next chapter and several others.

CHAPTER 10

Music Changes Us

"Music can change the world because it can change people."
Bono
Musician

"Music produces a kind of pleasure which human nature cannot do without."
Confucius
Philosopher

"When I hear music, I fear no danger. I am invulnerable. I see no foe."
Henry David Thoreau
Essayist, Poet, Philosopher

Quintessential scientific genius Albert Einstein was also a man of considerable compassion and numerous friendships, as displayed in the content of his voluminous correspondence and public speeches. One of his close friends was Queen Elizabeth of Belgium. In 1934, with political unrest and foreboding palpably present throughout Europe, Einstein learned that Elizabeth had recently lost both her husband, King Albert, and her beloved daughter-in-law to death. He quickly wrote a letter of condolence to Elizabeth, attempting to provide whatever consolation that he could. In addition to several other insightful comments, Einstein observed that life continues to unfold despite our losses and that indeed part of our consolation is "Mozart remains as beautiful and tender as he always was and always will be." Music, for Einstein, for Elizabeth, and for many of us, possesses a unique power of solace, consolation and comfort. It changes us.

But why and how can this be?

Before asking what the sciences of the psychology and neurobiology of music are showing us, let's pause and ponder these questions from **our own personal** experience.

Please take a moment with me, right now, to think about those occasions in your own life when music profoundly changed your mood, improved your attitude, enhanced your sense of well-being, bonded you to a loved one or to a group, made you literally ecstatic, gave you "goose bumps," comforted you, gave you courage, helped you to feel that life was worth living despite its many challenges, calmed you, energized you, made you want to move or to dance, made you feel romantic, made you contemplative, made you rejoice, or made you sad. Was it a particular type of music? A particular popular song, a passage in a classical work, a chord progression in jazz, an intense rhythm in rock music, or something else that produced such an emotional change in you? Did it occur only during a particular circumstance in your life? Or does it occur now with a particular melody, perhaps one that triggers a powerful memory? What is it? Can you define it? Or is it indefinable and completely ineffable for you—mysterious, but nonetheless wonderful? Or perhaps you are a fortunate person who can clearly identify and articulate precisely what seems "to do it" for you?

Some people routinely and reliably experience what the French call *frisson* (free-SÂN) when they hear certain musical passages. Roughly translated, *frisson* means "shiver" or "chills." Oftentimes, this sensation of being thrilled by a musical passage is accompanied by goose bumps (piloerection). Interestingly, the response can be partially blocked by a drug called naloxone, which antagonizes the effects of opioids, thereby providing a clue that the response is mediated by an increase in the endogenous opioid peptides in our brains that are also involved in other types of pleasurable experiences.

The Psychology of Music[1]

Studies on the psychology of music, its perception, its enjoyment, its performance, improvisation, and composition have produced an immense literature. Prior to the last century, however, much of this literature was reflective rather than empirical. Those interested in these topics **thought** about them, rather than performed actual experiments with human subjects other than themselves. Such speculation and conjecture are not without value but in recent decades, science has advanced considerably by using new tools and a more rigorous, quantitative approach.

Psychologists interested in the response of humans (and sometimes other animals) to music have been aided significantly in their studies by the brain imaging technologies described in Chapter 9. In addition, they continue to rely heavily on self-reports from carefully designed experiments with musicians and non-musicians. We began our own discussion here by asking you to consider the ways that music changes you and suggested that you examine and investigate your own responses to music. While each of us is undoubtedly unique in this regard, psychologists have found certain trends, correlations, and generalities that are at a level of **statistical significance**, that is, they occur more often than chance alone would predict.

Psychological studies have revealed that people listen to music for different reasons at different times. Chief among these purposes is to take advantage of music's ability to change mood. This is a powerful effect. It's often divided into two broad categories of change: (1) music to stimulate and to arouse and (2) music to sedate and to calm. This categorization may seem simplistic or even obvious but it certainly points to a commonly held experience of many people in many cultures.

For example, music that stimulates and arouses, including musical forms such as marches, national anthems, ceremonial prologues, patriotic songs, many operatic overtures, military drum cadences, many religious hymns, most rock music, dance music ranging from folk dances to disco, and most forms of music used in advertising, all tend to energize, motivate, and inspire the listener to **do** something. The "something" could be to display loyalty to one's country, to become excited about a sporting event, to win a war, to purchase something, or to dance until the nightclub closes. The composers and performers of such music, as well as those who utilize the music, take advantage of its power to change mood and thereby to induce behavioral change in the listener.

On the other hand, music can certainly sedate and calm, an effect seemingly opposite to arousal, but our brains are nonetheless aroused, though in a different way. Such forms of music include funeral music, "sad" popular songs that deal with loss or unrequited love, some forms of religious music, lullabies, music used as underscoring for "sad" scenes in films or plays, many romantic songs whether popular or classical, and music used as prologue for a solemn occasion such as a memorial. This type of music seems to cause us to be serious, attentive, consoled, calm, quiet, contemplative, and often sad as well.

Several characteristic parameters correlate highly in defining these two types of music. Stimulating music tends 1) to have a fast, steady, invariant tempo (hence the term "upbeat"), 2) to be relatively loud and frequently percussive, and 3) to be written in a major key, at least in the Western European tradition. (More about cross-cultural exceptions to this later.) Calming music tends 1) to have a slow but somewhat flexible tempo and takes frequent advantage of expressive tempo changes indicated, for example, as *ritardando* (deliberately slower), *rallentando* (slowing and dying away) or *rubato* (freely in time, stealing time from one note and giving it to another). It also tends 2) to be softer in dynamic and more *legato* (smooth) and 3) is often written in a minor key, or at least with the abundant use of minor chord progressions. Again, all are the conventions of the Western tradition. These findings are well documented in the psychological literature and you, yourself, will probably confirm that they are consistent with your own experience as well.

We humans seem to prefer a range of mood changes induced by music. For this reason, lengthy classical musical forms, such as the symphony or concerto, nearly always alternate fast and slow movements so that listeners experience a range of emotion and are therefore less likely to become bored. For the same effect, jazz performers will often alternate faster and slower songs (ballads) in their "sets" in a club.

Of course, psychologists have asked: Why do humans desire to listen to "sad music" when we are sad or distressed? Why not "happy music" to change our mood to a happier state and to relieve the pain of sadness, loss, or other apparently unpleasant feeling? After all, "happy music" can indeed make people happier.

The best explanation seems to be that we listen to "sad music" when we are sad because it is **coherent** with what we are feeling at the time. Such music apparently confirms to us that other people have also experienced what we are feeling and that we are therefore not alone in our sadness. The music assures us that other people understand us because they, too, have experienced similar pain. Sad music may also work, perhaps unexpectedly or paradoxically, to relieve our sadness by helping us to grieve a loss, to express our emotion physically (for example, by weeping), or to clarify and process our own pain in our minds so that we can move forward beyond the sadness.

Our common experience teaches us that music and movement are intimately associated, and as we observed before, the auditory cortex and the sensorimotor cortex of the brain are closely connected. This is not at all surprising, given the importance of the pulse in music (Chapter 3). Some types of music are quite literally designed to make us move—to make us dance or march, for example. It may be that dance music, emphasizing percussion instruments and a prominent, constant, persistent beat, was the first form of instrumental music that humans created (Chapter 13). Of course, synchronous movement binds the individual members of a group together, creating group cohesion, thereby changing mood. Think of soldiers marching together or a crowd of dancers in a disco, all moving rhythmically and synchronously together. Humans seem to find this coordinated, entrained, inter-twined movement to be highly pleasurable. Perhaps we feel less alone?

Another attribute of music is that it seems to help us to define ourselves, to find our identity with regard to our group. Humans are a highly social primate species. We rarely function well alone. But within our group, we constantly seek to define our role, our place, including our romantic attractions. Especially at certain stages of life, such as in adolescence or early adulthood, music seems to provide help with this process for many of us. Various studies have confirmed that the music of adolescence and early adulthood makes a lasting impression on us, so much so that in late life, this music still triggers powerful memories. [Aside: My own anecdotal experience in performing for the residents of nursing homes, many of whom suffer with various dementias, is that the popular music of their early adulthood, in this case, the Big Band Era, remains emotionally evocative for them, often eliciting latent memories.]

Finally, listening to music quite simply gives us pleasure. As we learned in Chapter 9, listening to music stimulates various parts of the limbic system, an evolutionarily ancient part of our brain intimately involved in reward and in the motivation to engage in specific behaviors. These rewards most often have to do with personal or species survival, such as the pleasure we obtain from consuming food or water, or having sex. Of course, various drugs such as alcohol, cocaine, nicotine, opioids, and others apparently activate the same neuronal circuits. These circuits and anatomical regions are also involved in mental illness including depression, anxiety, and psychosis, and in certain addictive behaviors such as compulsive gambling.

Fairly recent studies have demonstrated clearly that listening to our favorite tunes can initiate a rush of the neurotransmitter dopamine in the frontal striatum of the brain, an area of the limbic system that anticipates rewards. That in turn is followed by a rise in dopamine levels in the rear striatum, part of the pleasure / reward experience center of the brain. Dopamine is sometimes called the "feel good" neurotransmitter. Many neurons in the limbic system are **dopaminergic** and various addictive drugs also cause such neurons to fire.

The Ian Drury song title "Sex, Drugs, and Rock 'n' Roll" comes to mind here. Indeed, since music undoubtedly provides most of us (though not all of us, by the way) with genuine pleasure, the question remains: Why should music, like drugs, also activate these **survival** circuits? At least with drugs, we understand that these foreign substances mimic the effects of naturally occurring neurologically active biochemicals in the brain. With compulsive gambling, the reward is money, tied to survival. So, why music? Why organized sound? We'll address this question a bit in Chapter 13 because the consideration of this question reveals so much about our musicality as a species.

Much of the evidence that underlies these conclusions was obtained from detailed surveys and experimentation with college students, a readily available resource for academic psychologists. While you may have been a college student at one time, and perhaps are a student even now, the choice of this experimental cohort for study may well not be representative of humanity in general. Nonetheless, voluminous evidence exists that (1) music can change our mood, (2) music is used deliberatively by humans either to change or to confirm our mood, (3) music induces us to move synchronously, which gives us pleasure and a sense of belonging and cohesion with the group, (4) music helps us to define who we are, how we fit into the group, and even the nature of our romantic attachments, and finally (5) music gives us pleasure, sometimes intensely. That's a lot for sound organized against time to do!

More Objective Neurological Measures about Music and the Brain

So far, we've focused on what might be called traditional psychology, which is based upon the self-reports of human subjects and the direct observation of the behavior of humans in response to particular stimuli or under particular life circumstances. While this remains a very useful approach, science has acquired several new tools that provide a more objective rather than merely a subjective measure of the changes wrought by music.

In Chapter 9 on "Our Musical Brain," we listed and described these powerful tools and learned that they could be used for two broad types of analysis. The first is **structural** examination of the brain: its anatomy and various pathologies of its anatomy such as tumor, stroke, congenital anomaly, etc. The second type of analysis is **functional**, wherein neuroscientists are able to answer questions about which parts of the brain participate in its many activities. We learned that the functional tools, especially fMRI and PET scans, have helped us to gain a great deal of evidence about the specific parts of the brain that participate in the various aspects of music-making and we summarized those findings. We noted that virtually the whole brain "lights up" with music, hence neuroscientists have found music to be a useful probe into the functioning of our brains.

The next question is: Does listening to music, music training, and / or music-making **change** our brains consistently and persistently, either structurally or functionally or both? We know that it changes **us** but does it change **our brains** too?

This is a very important question because, in the last few decades, voluminous evidence has accumulated that the brain is **plastic**. Our brain changes with the demands placed upon it. These data come from both animal and human studies and they are compelling. While such plasticity is far greater in younger persons, it continues even into old age. Our brain responds no matter how old we are!

If once again we think for a moment about trying to determine whether the brain changes in response to music or music making, we could:

1. Dissect the brain post-mortem or structurally image the brain in the living person and look for changes at the "gross anatomical" level;

2. Study persons who have suffered a stroke, a tumor, or have a congenital anomaly that affects their ability to appreciate or to make music;
3. Examine the brain microscopically post-mortem and try to find structural cellular changes, say in musicians versus non-musicians;
4. Examine the brain biochemically and molecularly, looking for differences between musicians and non-musicians;
5. Measure persistent changes in the electrical impulses or the magnetic fields generated by those electrical impulses in various areas of the brain.

Neuroscientists have employed all of these approaches. In recent decades, they have documented music-related **structural** changes in our brains and these changes seem consistent with the changes that we might expect from the **functional** activities of various parts of the brain listed in Chapter 9.

One of the most well documented structural changes is in the size of the corpus callosum, the white matter that connects the two hemispheres of our brains. This is enlarged in the brains of musicians, especially in keyboard players, as we noted in Chapter 9. Because the two hemispheres of our brains divide the work of motor control, with the right hemisphere controlling the left side of our bodies, and the left controlling the right side, and because various aspects of music listening are also processed with lateral differences, musicians might well require greater "bandwidth" or more efficient communication between the two hemispheres. Hence, an enlargement in the corpus callosum "cable" connection might be expected.

Neuroscientists have also found differences in the gray matter of the brain related to music. For example, they've observed enlargement of the primary motor cortex, the somatosensory cortex, the premotor cortex, parts of the parietal cortex, and specific parts of the temporal cortex in musicians compared with non-musicians. In other words, those parts of the brain that are most intimately involved in music making may actually change physically in response to the demands of making music.

All of this seems reasonable but remember that in science, we must always be skeptical. These results remain provisional. Various studies have used different definitions of "musician" and "non-musician" and the experience and skill of the participants has varied. The conclusions from all such studies rely upon careful statistical analysis and such analysis in turn relies upon the sample size and many other factors that neuroscientists cannot always control.

One question uniformly pertinent to these studies is whether musicians become musicians because their brains are different, or whether their brains are different because they make music. This is the classic genetics versus environment or nature versus nurture question, which is reason enough for skepticism and caution in interpreting these results.

While the functional changes shown by PET and fMRI imaging must also remain provisional at this point, they are interesting. For example, fMRI has shown that musicians apparently show a greater response to sad music than do non-musicians, that is, musicians seem to be more emotionally responsive to music. Again, is this because musicians inherit a greater tendency toward musical responsiveness or do they develop it? And if they inherit this emotional responsiveness, is that part of the reason that they become musicians?

The limitation of both PET and fMRI imaging to reveal functional changes in our musical brains is that they operate over relatively longer intervals of seconds or tens of seconds, whereas the brain responds to music in the tens of milliseconds range. That's at least a 1000-fold difference. Fortunately, two other methods, EEG (electroencephalography) and MEG (magnetoencephalography), do measure changes occurring in the millisecond range, and while neither has the spatial specificity or resolution of PET or fMRI imaging, these two methods provide temporal information that is undoubtedly much closer to the timescale of neuronal firing induced by music in the brain. Each imaging tool therefore has its particular advantages and disadvantages. PET and fMRI have helped us to understand the anatomical areas of the brain involved in music-making and perception, whereas EEG and MEG help us to understand the temporal firing of neurons in response to music, with measurements in the millisecond range.

Summarizing:

Tool	Application
PET* and fMRI scans	Detailed information about anatomical areas involved but poor temporal correlation
EEG and MEG measurements	Limited information about anatomy, but much more accurate information about neuronal firing

*One other disadvantage of PET is that it requires a small amount of injected radioactivity, whereas the other three are completely non-invasive.

Before discussing the findings from studies with EEG and MEG, we'll need to learn a bit more about these two tools and the kinds of data derived from them. A salient difference is that EEG measures **electrical activity** whereas MEG measures **magnetic fields generated by electrical activity** in the brain. A common characteristic for both is that neuroscientists and psychologists must deal with something called **signal to noise ratio**, which is a measure of the strength of the signal (the change in activity directly related to the stimulus) compared with the background electrical or magnetic activity (the noise). This ratio is very low for both tools. Therefore, many repetitions (or trials) must be made in each experiment for each subject. This in turn necessitates careful work that is expensive to conduct. Nonetheless, the results are worth it because of the tremendous advantage of obtaining information about neuronal firing in response to a stimulus in real time.

The next step in learning about EEG and MEG is to describe how the data from these studies are represented. Generally, the results are expressed graphically to display the change in polarity or field (positive or negative) against time. An important concept used by neuroscientists is the **Event-Related Potential** or **ERP.** An ERP is simply a measured brain response recorded by the EEG that is elicited as the direct result of a specific sensory, cognitive, or motor event. Obviously, here, we're most interested in ERP's elicited by listening to music or by making music. The equivalent for MEG is the **Event-Related Field** or **ERF** (so named because of the magnetic field generated). In essence, then, neuroscientists and psychologists use these terms to describe **changes** in either the electrical activity or the magnetic field of particular areas of the brain in response to a specific stimulus, such as musical sound.

Neuroscientists have identified several different kinds of ERP's and ERF's, categorizing them as part of their analysis. These categories of ERP's and ERF's are assigned a variety of names and abbreviations that enable neuroscientists to communicate more easily with each other about the changes and to write about them more succinctly in the scientific literature. I've made a deliberate choice in this book **not** to discuss the many types and subtypes of ERP's and ERF's. Frankly, such detailed scientific nomenclature adds little at our level of discussion and doesn't contribute much to our purpose here. But the reader should be aware that this is a complex scientific field of inquiry, constantly increasing in complexity as more and more observations accumulate and as they are collected, collated, categorized, and analyzed.

Musical Syntax[2]

One concept that is definitely necessary for us to understand the response of the brain to music as measured using EEG or MEG is **musical syntax.** Just as languages have syntax, the "rules" of the language that guide the speaker or writer in putting words together, music has syntax as well. We've discussed some of these rules and conventions in Chapters 5 ("Melody") and 6 ("Harmony"). As you recall from those chapters, in the Western tradition, the musical key (the particular diatonic scale being used for a melody) and the chord progression combine to create expectations. For example, we expect the melody to end on the tonic note of the scale and we expect the chord progression (in tonal music) to move ultimately toward the tonic chord (I). Any note not part of that diatonic scale (that key) or chord progressions not moving as expected will create a sense that the syntax of the music is incorrect.

It's a bit like putting the verb at the **end** of an English sentence, rather than after the subject and before the object. In classical Latin, such word placement (subject-object-verb) is perfectly syntactical, but not in English. The violation of syntax is used to good advantage and effect in the dialogue written for the Yoda character in the *Star Wars* movies as he says: "The Force with us is." Such dialogue makes Yoda's expressions different and exotic, precisely the purpose of the screenwriter.

Western listeners can perceive a wide range of degree of "violations" of syntax, melodic, harmonic, or rhythmic, which composers use to good advantage as they establish expectations and resolutions. In other words, some syntactical violations would be perceived immediately as "wrong," whereas others would be much less surprising or bothersome to the listener, and are likely to add interest to the music. Musicians and other experienced listeners perceive finer degrees of syntactical differences than non-musicians. Using various techniques, composers are also able to prepare the listener for a violation of syntax. We'll discuss this in greater detail in Chapter 12, "The Composer's Toolkit."

Musical syntax and its violation have been very valuable in studying the response of the brain to music as measured employing EEG and MEG–even in studying the brains of very young children because both tools are non-invasive and without the risks of X-radiation or radioactivity. The comparison of the EEG or MEG signals generated by musicians compared with those from the brains of non-musicians has been especially revealing for neuroscientists.

Let's discuss one representative type of experiment that utilizes musical syntax to probe the brains of participants as they listen to music. In this case, investigators studied the participants' responses to three different progressions of five chords each:

Progression (a): I – I (*first inversion*) – IV – V^7 – I

This is a very standard progression that begins and ends with the tonic, with the V^7 as the leading chord.

In C-major, the chords would be:

I = C+C'+E'+G'

I (*first inversion*) = E+C'+G'+C"

IV = F+C'+F'+A'

V^7 = G+B+D'+F'

Progression (b): I – I (*first inversion*) – ♭II^6 – V^7 – I

In C-major, the ♭II^6 chord, also known as the Neapolitan 6th chord, can be played:

♭II^6 = F+D♭+F'+A♭'

Recall that the usual ii chord is a minor chord played, as a triad in C-major, D+F+A. Therefore the ♭II^6 is not part of the usual syntax of this scale or the chord progressions derived from it. That said, it is a perfectly consonant chord (actually simply a D-flat major chord) and in the key of D-flat major, it would be the tonic.

Progression (c): I – I (*first inversion*) – IV – V^7 - ♭II^6

Progression (c) seriously violates the expected syntax of tonal music because it ends with a leading chord that "demands" resolution to the tonic major chord. The key was established syntactically as C-major with the first two chords of the progression. The ♭II^6 chord is not part of the key structure at all. Worse, it's a leading chord in its own right and the progression is therefore ending with a leading chord, something just not done conventionally in Western music. Everything is wrong, accept that the ♭II^6 is consonant, which ensures that the brain's perception is not due to dissonance, but rather only to the type of chord and where it's placed in the progression.

This type of experiment is cleverly design because the chord progression is short, simple, and the ♭II^6 chord is perfectly consonant, but not at all musically syntactical in this progression. Neuroscientists use a wide range of controls to avoid artifacts in their experimental results. Recall that artifacts are experimental data generated from a quirk in the experimental design as opposed to the defined stimulus under investigation. Remember: Scientists must always be skeptical, especially of their own data.

This study is representative of a class of similarly designed studies which, taken together, document quite compelling evidence for the following findings:

1. The ERP for Progression (c) when ♭II^6 sounds is of much larger magnitude than for Progression (b), which is in turn greater than the ERP for Progression (a). This result shows that the highly unusual placement of a non-syntactical, irregular chord in the progression especially at the end of the progression, where it is least expected, trig-

gers specific, detectable electrical activity in the brains of the participants. Moreover, this response is found specifically in the right anterior (frontal) cortex.
2. Non-syntactical notes in a melody also elicit ERP's.
3. These responses are rapid, in the range of less than 100 milliseconds.
4. Musicians respond significantly more than non-musicians to incongruous, non-syntactical pitches or chords, especially as measured in their frontal and temporal lobes.
5. Trumpet players show larger signals with trumpet sounds, whereas violinists show larger signals with violin sounds.
6. Males show a greater degree of lateral localization of the response than do females.
7. Musicians seem to encode pitches better than non-musicians.
8. ERP's occur with non-syntactical musical patterns even if the participant is not listening specifically for it.
9. Even a small amount of musical training changes the response of the brain to violations of ordinary musical syntax.

The last finding is very revealing and important. Musical training apparently changes the response of the brain to music. This is important because the finding indicates strongly that the brain changes in response to music-making. This would indicate that environment / nurture may well be the cause of the change rather than some kind of strictly inherited, genetically determined, pre-existing difference between musicians and non-musicians. Neuroscientists readily acknowledge that we need many more **longitudinal studies**, where there are appropriate "treatment" and control groups studied over time. We'll discuss some of these studies in Chapter 14, "Our Musical Children."

By the way, the idea is **not** that musicians are somehow "better" than non-musicians, though indeed musicians may preferentially have developed desirable skills. Rather, the idea from these studies is that the brain apparently **changes** when demands are placed upon it, and neuroscientists have found nothing more demanding of the capabilities of the human brain than musical performance because it requires the entire fully engaged brain, apparently with many of the brain's circuits activated.

In summary, then, music changes us and this change seems to be reflected in our felt emotions and also physically, including alterations in both the structure and the functioning of our brains in highly specific, empirically measurable ways. Whether we inherit a greater or lesser propensity or ability to experience such change, and to what degree, remains a valid question. As always, these results are provisional, requiring much more replication and documentation, especially with longitudinal studies employing control and experimental groups.

But, if we assume that music does change the structure and functioning of our brains, can we use music **therapeutically**? Can we use it for healing and for improving our well-being in clinical practice? That's the subject of Chapter 11.

CHAPTER 11

Music Heals Us

"Music is a therapy. It is a communication far more powerful than words, far more immediate, far more efficient."
Yehudi Menuhin
Violinist, Conductor

"I think music in itself is healing."
Billy Joel
Singer, Songwriter, Pianist

"One good thing about music is when it hits you, you feel no pain."
Bob Marley
Singer, Songwriter

Denny, my friend of 25 years, developed Parkinson's disease relatively early in his life, when he was only about 50 years old. Parkinson's disease can be devastating, in part because it is a disease of the brain. Affecting one million Americans today, the initial symptoms usually include tremor, bodily rigidity, and slow movement. Later, depression, dementia, and more serious motor deficits, including vocal and breathing symptoms, often occur. While the progression of the disease is inexorable, fortunately with the advent of levodopa and other dopaminergic drugs, the prognosis isn't nearly as grave as in the past. Neuroscientists have obtained strong evidence that cells in the substantia nigra (part of the basal ganglia), a region of the midbrain, die for reasons not fully understood and that this cellular death results in reduced production of the neurotransmitter dopamine, which is required to regulate normal movement.

Nonetheless, Denny, like so many who suffer with Parkinson's disease, is resourceful, courageous, and determined. He found, on his own, that participating in ballroom dancing is helpful to him. It helps to normalize his movement. In other words, moving to music is definitely therapeutic for him.

Denny is not alone in his observation of the benefits to be derived from music. A recent report[1] shows that singing is therapeutic for people with Parkinson's disease. After two months of singing twice per week, vocal quality, including control of swallowing—often impaired by the disease—was improved in these patients. In addition, group singing enhanced social interaction.

Just as Denny discovered the power of music experientially for himself with his Parkinson's disease, neuroscientists have shown empirically that entrainment to a pulse, to rhythm, aids those with a variety of other motor-related brain diseases, including patients with certain kinds of strokes, children with cerebral palsy, patients suffering from traumatic brain injury, and patients with Huntington's disease. What may happen in each of these cases is that the musical pulse provides the brain with a rhythmic center, a "cue," a "hook," to entrain and thereby to improve movement, even if the part of the brain that would normally regulate that particular movement sequence is damaged or inoperative. We've previously observed (Chapters 2, 3, 9, and 10) that music and movement are intimately interconnected and that the pulse is the foundational organizing principle of music, cross-culturally.

Music as Brain Therapy[2,3]

The notion of using music therapeutically has been extant in humanity for many thousands of years. The ancient Greeks considered music to be an important part of medicine, using it even for the curing of hangovers after nights of heavy drinking. The ancient Jewish people documented the therapeutic use of music in their scriptural texts. Native American peoples regarded music as part of their arsenal of tools for healing. In more recent times, music has been used for its aesthetic and emotionally evocative properties as an "honorable mention" for the healing arts—helpful but ancillary. Even the United States military employed musicians to help wounded soldiers. Since about 1990, however, music has attained a much more important position in clinical medicine as a result of progress in neuroscience, which in turn was enabled by the imaging technologies that we've discussed and described in Chapters 9 and 10. This was a step-change, showing clinicians that the benefits of music could actually be traced to biological alterations in the patient, just as drugs or surgery cause demonstrable physiological alterations valuable to healing. In other words, clinicians could postulate a **biological mechanism** for the benefits of music, making it a rational and even an orthodox therapeutic modality.[4]

The range of such physiological changes is large and well documented.[5] Perhaps you can think immediately of changes that might happen therapeutically based on everything that you know about music at this point. One that immediately may come to your mind is in the functioning of the heart. After all, in Chapter 3, we discussed the pulse and used the heartbeat as an example of the biological underpinning of our human perception of the pulse. In fact, it's known that listening to appropriate music can reduce heart rate and blood pressure, both the systolic (the upper number) and diastolic (the lower number) pressures. A favorable effect of music is also observed in postsurgical cardiac patients, where music has been shown to reduce cardiac complications. Anxiety and respiratory rate were reduced in patients who listened to calming music.

Less obvious are the effects of music on various hormone levels in the body. For example, cortisol is a steroid hormone produced by the adrenal glands in response to physiological stress of any kind, whether the stress is biological (surgery, trauma) or emotional (a difficult life circumstance, grief, separation, anxiety). [Aside: the secretion of cortisol is in turn regulated by two other hormones, adrenocorticotrophic hormone, produced in the pituitary, and corticotropin- releasing hormone, produced by the hypothalamus of the brain. Thus, the level of cortisol in the blood is regulated ultimately by cells in the brain.]

Cortisol is a two-edged sword. On the one hand, it reduces inappropriate inflammation and prepares the body to heal itself. On the other hand, it can weaken the immune system, delay wound healing, and cause bone loss. Cortisol is elevated for short periods of time to help the body deal with an immediate, potentially life-threatening event. Cortisol blood levels are therefore tightly regulated to obtain its benefits as an immune system modulator, preventing both the over activity of the immune system and counter-productive inflammation. However, when it is chronically secreted by the adrenal glands at a higher than normal level over a long period of time due to chronic stress, cortisol can be deleterious. Many scientists think that high cortisol levels thereby contribute to stress-related diseases.

In any case, as shown in various studies, when patients listened to the music of their choice, cortisol levels were reduced during medical procedures. This indicates that the patients' overall anxiety was decreased and that their basic physiological functioning was improved by music. During surgical procedures, the amount of anesthetic required is reduced. In other words, music is an anxiolytic, and a powerful one at that.

Endorphins, which constitute a class of three peptide hormones produced in the brain and pituitary gland, derive their name from a contraction of two words, namely, "**endo**genous" and "mo**rphine**." Though produced physiologically, they act very similarly to morphine and synthetic opioids in relieving pain and in producing a form of euphoria. Endorphin release is apparently responsible for the "runner's high" greatly prized by long distance runners. Endorphins bind to the same receptors on neurons as bind opioid drugs. Just as cortisol helps the body to deal with the immediate consequences of trauma, the endorphins aid with the regulation of the immediate pain following injury that might be counterproductive to survival.

In any case, music seems to stimulate endorphin release, at least in some people under certain situations. The evidence for this comes from the ability of naloxone, an opiate antagonist, to reduce the euphoria derived from music. You may be aware that naloxone can be life-saving in patients who have overdosed on opioids, including fentanyl, because it blocks the binding of the opioid to its receptor in the brain. Thus, it would appear that part of our euphoria derived from music, just as with the euphoria derived from intense aerobic exercise such as long-distance running, is mediated through the endorphin system.

As we mentioned, cortisol can decrease the functioning of the immune system. Under chronic stress, cortisol levels may reach inappropriately high levels for long periods of time, reducing the effectiveness of the immune system, which itself is a very complicated and critical system that employs scores of different chemical signals and mediators in its regulation and relies on many different types of cells, all hopefully working together flawlessly to protect us from infectious invaders. Immunologists can measure the activity of the immune system by measuring the blood levels of these various chemical mediators. The mediators include proteins such as interferons, interleukins, and others. Data are now accumulating that listening to music may well "normalize" the functioning of the immune system so that it does its job more effectively, without being overactive or underactive.

Therapeutic Areas

In terms of specific therapeutic goals, the best evidence for the effectiveness of music is in four areas:

1. Improvement in motor function and control;
2. Improvement in cognitive function, especially attention and memory;
3. Improvement in speech and language function;
4. Improvement in psychological function, regardless of the disease, especially the reduction of anxiety and enhanced socialization with consequent improvement in quality of life.

All four of these involve the brain, and while the rest of the body certainly also benefits, especially with improved motor control, we could well regard music as powerful brain therapy, useful by itself, and also synergistically with other therapies as well. Music enhances well-being in the absence of disease, as we learned in Chapter 10, but it also treats specific brain functions affected by disease.

We've already touched on the assistance that music can provide to patients with various motor deficits. In patients with damage to specific motor areas of the brain, cuing with a gradually faster external tempo seems to be helpful, perhaps because it "pushes" the capabilities of the brain. What seems to happen, based upon imaging, is that intact areas of the brain, those not damaged, take on functional responsibility for movement. This concept recapitulates our discussion in Chapter 10 about neuroplasticity. From all the data that we have, the best explanation is that the brain can change itself to overcome the deficit if assisted by the exogenous pulse.

Stroke, and other brain injury such as vascular dementia, as well as developmental abnormalities, can cause serious deficits in many different aspects of cognitive functioning, including memory, planning, and personal interactions. Alzheimer's disease is, of course, the condition that receives much attention now, in part because it accounts for 60 to 70% of all dementias and affects an estimated 5.5 million Americans. Music therapy can be very useful here, helping patients to recover apparently lost memories, calming agitation and improving socialization. It would appear that musical memories persist well after other intellectual capabilities are lost. This observation also provides support to the notion that musical expression is at the core of our humanity.

Children on the autism spectrum or with other, often congenital, behavioral challenges can also be helped greatly by music. Case reports in the literature show that when such children are provided with intensive exposure to music, they can recover a sense of the "musical self" and thereby gain a sense of competence and confidence that they did not have before. Often, music therapy is used in conjunction with traditional psychotherapy and psychiatric drugs to assist the children as they deal with the brain disease that leads to their "acting out" behaviors or other inappropriate actions. Here again, music functions as an external cue, something that they can apparently use to organize behavior and attention, expanding their repertoire of desirable behaviors.

Music-making can even **prevent** the changes due to normal aging. This may seem remarkable but, based upon our observations in Chapters 9 and 10 that music-making apparently activates the entire brain and provides a significant challenge to our cognitive capabilities, it is not entirely surprising. If the brain is plastic, as it certainly appears to be, and if it can change both structurally and functionally as a result of an extraordinary demand, such as making music, why wouldn't music-making encourage the development of auxiliary capacity that could be useful as a reserve in the event of damage to the brain?

While the evidence supporting this notion is not yet abundant, the results are intriguing. For example, aging adult musicians apparently maintain speech fluency, memory, attention, and rapid brain processing better than their age-matched, non-musician peers. Because music-making requires the use of all of these abilities, or abilities closely related to them at a high level, the rationale for their maintenance in older musicians seems clear. Many more studies are required to validate these findings, however, to obtain sufficiently compelling data.

It may surprise you that adult musicians appear to maintain speech fluency more readily than their peers. Consistent with this finding is the use of music therapy in aphasic stroke patients. Aphasia is simply the loss of the ability to understand or to express speech as a result of damage to the areas of the brain responsible for those functions. Such deficit may or may not be accompanied by the loss of the ability to read or to write or both. All of these deficits can occur to different degrees of pathology. A patient may have complete loss of speech, for example, or only a loss of some fluency in his or her native language. As we shall discuss in more detail in Chapters 13 and 14, much evidence exists that language and music are closely intertwined, both in human evolution and in the neurological structures responsible for each of them.

One very common observation is that non-fluent aphasics may be able to sing words that they cannot speak. In other words, they are able to form the word perfectly fluently if the brain considers it to be musical and vocalizes it with pitch and rhythm. This indicates strongly that the areas of the brain responsible for speaking a word and singing a word differ. The phenomenon has become a staple principle in the therapy of such aphasics. The patient sings short phrases with the music therapist and as the patient progresses, the sung phrases become longer. Gradually, new circuits are created and new, undamaged areas of the brain are recruited for speech. In essence, if the speech centers of the left hemisphere are damaged by stroke, carefully directed singing, in pitch and rhythm, apparently recruits areas of the right hemisphere so that language fluency becomes much more proficient. Incidentally, this change is confirmed in results from imaging studies including MRI and EEG.

Methods

Finally, music therapists categorize their methods into four broad categories:

1. **Musical Improvisation.** The client makes **original** music, either with the voice or often with a simple instrument. The quality of the music is not important. Under the

guidance of the therapist, the client is encouraged to express herself or himself musically. This has been helpful in many different psychological disorders ranging from the treatment of autistic, non-verbal children, to aggressive adolescents, to psychotic adults.

2. **Making Music Composed by Others.** The client employs either the voice or a musical instrument, often together with other performers, to recreate the music. This is very helpful for motor development and where the necessary intellectual capability is present, for learning and translating musical notation.
3. **Listening to Music.** The client attentively listens, often to peaceful and calming music. This is helpful in improving mood, reducing anxiety, and directing attention. Alternatively, of course, the music can be used to entrain desired movement.
4. **Combination Therapies.** Here, a cocktail of specifically designed therapies is employed, tailored to the client. Perhaps cognitive therapy, physical therapy and / or behavioral therapies are combined with music therapy to achieve the therapeutic goal. No one therapy is sufficient by itself, but together they are synergistic and much more helpful to the client.

Music therapy is a relatively recently accepted and rapidly evolving modality of the healing arts. As we learn more about the underlying mechanisms and rationale of the therapy, it will become more and more effective. Neuroscientists and music therapists are already learning much from each other and working together to advance their service to patients.

With that, we've completed three chapters that have dealt primarily with our musical brain. We've dealt with the basic musical neuroscience of the brain in Chapter 9, the psychology of music in Chapter 10, and the use of music therapeutically here in Chapter 11.

Now, let's move on to a new line of inquiry: musical composition, the art and the science.

Like drama, music relies both on writer (composer) **and** performer (actor, musician), who may be the same person(s), but very often isn't. While some music may never be written down, it is always "composed," even if the composition is improvisatory and transitory with immediate, real-time performance. We've discussed musicians and musical performance at length in Chapter 7. But what is the role of the composer? How does the composer use what we know about musical perception, either deliberatively or with intrinsic, unconscious capability? What are the "tricks" of writing emotionally evocative music? Now, it's time to answer those fascinating questions, which reflect so much about the way that music really works. Together, we're going to investigate the composer's amazingly powerful toolkit.

CHAPTER 12

The Composer's Toolkit

"Without craftsmanship, inspiration is a mere reed shaken in the wind."
Johannes Brahms
Composer

"The old idea of a composer suddenly having a terrific idea and sitting up all night to write it is nonsense. Nighttime is for sleeping."
Benjamin Britten
Composer

"If a composer is to reach his audience emotionally—and surely that's what theatre music is all about—he must reach people through sounds they can relate to."
Richard Rogers
Composer

"If a composer could say what he had to say in words he would not bother trying to say it in music."
Gustav Mahler
Composer

"If anyone tells you [that] classical music is dead in the 21st century, all it means is that they don't go to the cinema."
Howard Goodall
Composer, Music Historian

It is rare indeed for any one person to make historic, highly memorable and even seminal contributions to more than one area of human endeavor especially if the disciplines are completely unrelated to each other. Leonardo da Vinci was a very rare genius who crossed over between science and the visual arts, making highly significant, even revolutionary contributions to both. Da Vinci is esteemed both as a renowned painter and as a prescient scientist / inventor.

Slightly more common are those who have contributed significantly in two or more areas related closely to each other. Sir Isaac Newton exhibited extraordinary genius in both mathematics and physics by inventing the calculus and by formulating the basic laws of motion, thereby becoming one of the founders of Western science. In the visual arts, Michelangelo was a painter, sculptor, and architect who created many internationally recognized and revered masterpieces in all three of these art forms.

Such geniuses are commonly described as polymaths. The power of their intellects is so great that no single discipline can contain it. Interestingly, only a few polymaths have crossed over between a scientific discipline of any sort and the composition of music to make lasting contributions to both. One possible contemporary example is Brian May (1947) who was composer for the rock band Queen ("We Will Rock You") and who is now the author of several books about astrophysics.

Alexander Borodin, Composer-Scientist

By far the most famous of these scientist-composer polymaths is the Russian composer and chemist, Alexander Borodin (1833 – 1887)[1], whose contributions in both areas demonstrably withstand the test of time. The listing of a composer's works as part of the standard orchestral repertoire—compositions performed widely and often by orchestras throughout the world—reveals the importance and accomplishment of that composer. Fewer than one hundred composers are represented on the standard list. Three of Borodin's works are frequently performed, namely, his tone poem *In the Steppes of Central Asia*, the *Polovetsian Dances* from his opera *Prince Igor*, and the *Symphony No. 2 in B minor*. Also impressive is that several themes from these three compositions, as well as from other works by Borodin, were used as the musical source material for the 1954 Tony Award-winning Broadway musical, *Kismet*. Memorable popular songs from this show, all based on Borodin's compositions, include "Stranger in Paradise," "And This is My Beloved," and "Baubles, Bangles, and Beads."

Even if Borodin had not been a renowned composer, he would have been remembered for his contributions to the science of chemistry. Perhaps surprisingly, Borodin seemed conflicted about the time that he took from his science to compose. As he remarked, "I am a composer in search of oblivion; and I'm always slightly ashamed to admit that I compose;"—and on another occasion—"Music is a pastime, a relaxation from more serious occupations." His humorous admonition was that "respectable people do not compose music—or make love—as a career." Evidently, he considered himself to be a scientist first and a composer second, and the latter only by avocation. Perhaps he failed to understand that some other scientist eventually would have discovered his chemistry, but no one else could ever have written his music.

Borodin was the illegitimate son of a Georgian prince (Luke Gedeoneshvilli) and his serving maid. As was common at that time, the infant son received the last name of one of the prince's serfs, and thus he became Alexander Borodin. Fortunately, Alexander received a fine education, becoming interested in both science and music early in his childhood. He was trained as a physician, then earned a doctorate in chemistry, and ultimately received a professorship in organic chemistry at the young age of thirty-one. Borodin was particularly known for his work with a class of compounds called aldehydes and published a considerable amount of novel chemistry, including the important "aldol condensation reaction," a chemical reaction between two carbon atoms necessary for our contemporary synthesis of atorvastatin (Lipitor®), the cholesterol-lowering drug, and many other important compounds.

Borodin had yet another passion—the academic advancement of women. He felt strongly that women ought to have opportunities equal to those of men in the sciences and became one of the founders of the School of Medicine for Women in St. Petersburg.

Unfortunately, his life was cut short by heart failure and he died suddenly at the age of 53. In those all too few years of life, he accomplished much as a composer, chemist, teacher, mentor, and pioneering feminist.

As with several other Russian composers of his time and later (Rimsky-Korsakov, Mussorgsky, Tchaikovsky, Rachmaninoff), Borodin was a melodist of the first rank. Perhaps more than any other quality, it is his melodic inventiveness that endears him to so many listeners. Certainly, the business-savvy producers who invested in *Kismet* understood the potential appeal of his melodies, even to 20th century American Broadway audiences. Of course, melody, as we've discussed (Chapter 5), is commonly one of the most important tools of the composer.

Improvisation = Composition

Regardless of the specific tools used to create music, **all** music is composed. Some of you will immediately rankle at this statement. You'll say: "What about jazz improvisation?" "What about non-Western music where there's no defined notation?" "What about **me** making up a lovely tune as I shower in the morning? Am I thereby a composer?"

Indeed, you are! Composition is ***not*** synonymous with writing music down on paper, recording it, or preserving it in any way. Musical improvisation is essentially musical composition, performed immediately by the composer, in real time, without delay.

In the Western tradition, we most often think of musical composition as something carefully thought out, reliant on this mysterious thing called "inspiration," and then written down so that it can be performed by other musicians, preserved "forever." We tend to think of musical improvisation, on the other hand, as being different, ephemeral, and not really composition. But it **is** composition, though often not recorded or able to be reproduced with fidelity.

Many great (written-down) composers have been fine improvisers. J. S. Bach is particularly known for his improvisatory abilities. In his time, any organist worth his pay was an excellent *ad hoc* improviser for church services. Bach had to be good at this.

Writing music in notation has both advantages and disadvantages. The chief advantages are (1) a composer who utilizes notation can consider a plethora of different ways for expressing novel musical ideas and often has a lot of time to think about them, experimenting before writing anything down in final form; (2) the ideas are preserved and can be performed by others, even by more than one hundred others simultaneously, as in a symphony orchestra; and (3) lengthy, complex compositions become much more easily created. The chief disadvantages are that (1) the inexact nature of musical notation—it's really a mere approximation—can never perfectly reflect the composer's intentions, thereby limiting the composer's expression in significant ways, and (2) the composer of necessity relies upon the **performer** (individual musician, group of musicians, conductor) to interpret the musical ideas encoded in this inexact, imprecise notation.

History provides prominent examples of upset or angry composers who thought that an uppity, arrogant conductor misinterpreted their well-crafted score. For example, Maurice Ravel was very upset with the famous conductor, Arturo Toscanini, for choosing what Ravel regarded as an inappropriately slow tempo for the very well received *Bolero* upon its American premiere. After the piece finished, the composer, who was in the audience, surrounded by approving applause resounding in the concert hall, was so angry that he refused to rise from his seat when Toscanini acknowledged him.

When Sergei Rachmaninoff was making the first electronic recording of his widely beloved *Piano Concerto No. 2* with himself as pianist and Leopold Stokowski as conductor, the two of them nearly came to physical blows over the interpretation. The contentiousness was so great that this first recording did not include the third movement of the concerto because the two musicians simply could not agree. Ultimately, a second attempt in 1929 did achieve common interpretive ground and an excellent restored version of this fine recording can be heard online.[2]

Musicians often improvise together, as if in conversation. In fact, many musicians find group improvisation to be one of the greatest, though certainly most demanding pleasures of making music. A common type of such musical conversation occurs when jazz musicians "trade fours." Recall that jazz often employs, as a starting point, the basic architecture of a particular popular song, with the song's undergirding chord progression and basic thematic material. This serves to provide some organization for both listener and performer. After each musician in the group has played an extended (12 to 16 bar) solo, the performers will sometimes repeatedly improvise 4 bars (or more), alternating between each other and the drummer for an extended period of time. Each 4-bar phrase will build sonically and coherently on the one before it, perhaps imitating it, perhaps adding a surprise change, perhaps deliberately making the same mistake or alteration in stating the organizing theme. Sometimes the musicians may move musically to an entirely different popular tune, even if briefly, and "talk" about that one for a while. The whole process is a delight for listener and performer alike in part because of its spontaneity. The music of other cultures is often equally spontaneous, and that will be a topic for Chapter 13.

Even in the case of written scores, as we saw with Ravel and Rachmaninoff above, there is opportunity for substantial variation in interpretation and in the ultimate performance. Symphony players report that some conductors are very rigid, however. Such conductors have one "perfect version" of the piece "in their heads" and it is simply not ever to be changed. Other conductors may "improvise" at least just a bit. If something happens in performance not quite like what happened in rehearsal (a different sonic balance, a soloist taking a different tempo) such a conductor will "go with the flow" and make music in the moment, in real time. She will easily, willingly, and deliberatively "adjust" to keep the music "alive."

The Toolkit[3]

Whether a composition is created through improvisation or by using careful notation transcribed with pen or computer, what are the tools that the composer can use? We've discussed many of these (Chapters 3, 4, 5, 6). They include **melody, tempo** (fast, moderate, slow), the **type of scale** (major or minor diatonic scales, full chromatic, pentatonic, whole tone, modal, cross-cultural, microtonal), **harmony** (choice of chord progression), **dynamics** (*fff, ff, f, mf, mp, p, pp, ppp*), **timbre** of the instruments used to perform the music and, of course, fundamental to all music, **rhythm** (the organization of the pulse, including both sound and silence).

In this chapter, we'll add some nuance to our understanding of the use of these critical, basic tools and describe the current scientific understanding about how these tools are used most effectively. Ultimately, composers want to stir some kind of emotional response in the listener and, as we've learned, scientific findings provide abundant verifiable evidence that music is extremely effective at achieving that result—documented both by psychological self-reports and with neuroimaging technology (Chapters 9, 10, 11). But achieving the emotional power of music is no mere accident. It requires skill, craft, inspiration, imagination, critical assessment and, frankly, hard work.

Perhaps the first question to consider is how composers do this work, especially the process that composers use to compose written music. From this point forward, we'll deal primarily with notated music, but many of the same principles apply to improvised music, especially for a musician who is highly trained and experienced in improvisation.

As we described in Chapter 10, imaging studies performed with musicians in the act of composition reveal that certain areas of the brain are activated, and interestingly, other areas are suppressed, as they compose. The areas deactivated during composition are also at rest during dreaming and meditative states. On the other hand, the areas that are activated also control language, sensorimotor skills, and the assessment of syntax.

Like any competent speaker or writer, the composer must decide the purpose of the composition and the audience to whom it is directed. When John Phillip Sousa wrote a military march, for example, his goal was **not** to compose a complex four-movement symphony or a romantic popular song. The purpose and the audience are very important determinants of the final form of the composition. Also, the composer must decide about the ensemble and instruments that will be performing her new work. Will it be a piece for unaccompanied violin, or church choir, or full symphony orchestra, or theater pit orchestra, or college wind band, or small chamber group? Is she writing on commission for a famous popular singer, hoping to make both of them rich by composing an immensely successful "platinum" single, or is she writing primarily because she is simply inspired to write, whether or not she can even find a publisher?

With those questions answered, composers use a wide variety of techniques to begin. While some composers work almost exclusively with their mind's ear, most seem to begin by improvising, a process often called "noodling," at a keyboard, experimenting with various ideas. When a "noodle," *e.g.*, a chord progression or a melody or even a simple bass line, appeals to them, they write it down. Even if they don't use it for the particular composition at hand, they may use it for something else altogether at a later date.

Some composers begin their composition with a chord progression, or even just a bass line that they like. This is often the case in certain popular genres where the chord progression is well established for a particular type of tune–the"blues" and so-called three-chord rock 'n' roll tunes are examples. Some composers always begin with a melody, or perhaps a 4- or 8-bar "motif," which is the beginning of the melody. Perhaps, they create this melody because they want it to complement a set of lyrics that has already been written by a collaborator, or that they, themselves, have written. This is an interesting point. Some composers are also brilliant lyricists (one thinks immediately of Cole Porter), whereas others always collaborate with a partner.

Gradually parts of the composition begin to take shape. If the composer has started with a melody, it must be harmonized with a chord progression that complements the melody. For centuries past, all of this was done at a keyboard, but now thanks to recently developed technology, it's done with computers equipped with specific software designed for composing music, such as *Finale*® or *Sibelius*®. This software is immensely useful to contemporary composers because they are not only able to work out the melody and harmony but can "play it back" using the computer and thereby obtain some notion of what the final version will sound like, adding all of the instrumental timbres that they plan to use. Before the advent of such software, of course, everything had to be done at the keyboard, or in the case of the gifted composer, entirely in his own "minds ear." Another advantage of the software is that it can print out the parts for the musicians. No more copying parts laboriously by hand from a score!

In any case, this first stage is highly akin to simple improvisation. The major difference is that the composer readily discards anything that doesn't "work" and can do so immediately. This enables the composer to go through thousands of musical ideas, keeping the ones he likes and discarding the duds. It also permits the production of a highly refined final score.

The ultimate goal is always emotional impact on the listener, which of course can take a wide range of forms, as we discussed at length in Chapter 10. The composer is fully aware of all of the psychological phenomena that we discovered in that chapter: slower tempi are perceived as sad, and faster tempi energize the listener; minor keys create a somber or sad mood, whereas major keys are usually "happy;" some chord progressions are within the expected syntax of the music, whereas others violate syntax, thereby surprising and perhaps even irritating the listener; tonal music requires a return to the tonic of the key for resolution, etc. In addition to this standard palette of techniques, composers have many other devices as well, often described—you guessed it—in Italian.

Some of the common devices that you can hear in most pieces of music include:

1. **Changes in tempo.** Composers can indicate a gradually increasing tempo with the term *accelerando* (gradually faster), *affrettando* (hurrying) or *stringendo* (pressing forward). Or they can indicate a slower tempo with terms like *rallentando* (gradually slower), *ritardando* (gradual slackening of the pace) or *allargando* (slower, broader,

and louder). Other terms are also useful such *meno mosso* (less motion) or *poco meno mosso* (a little less motion) or *poco a poco* (little by little) or *a tempo* (a return to the original tempo).

2. **Changes in dynamic.** Composers often indicate a sudden change in dynamic with the word *subito* (suddenly, immediately) followed by the new dynamic marking. A more gradual increase in loudness is indicated by the word *crescendo* and a gradual decrease in loudness by the words *decrescendo, diminuendo*, or *morendo* (literally dying away).
3. **Changes in key.** This is a very useful device to create dramatic impact, maintain interest, or introduce new thematic material. Key changes are often used in popular songs toward the end of the piece and there the key change is almost always **up** by at least a half step and often more than a half step to confer a new sense of energy to the composition. Classical composers use key changes as part of developing thematic material.
4. **Refinement of the sound and rhythm.** Composers utilize many words and symbols to help the performer to produce the sound and precise rhythm desired. For example, the composer may write *dolce* when he wants a sweet, delicate performance, *espressivo* when he wants the sound to be expressive and evocative, *rubato* when he wants the performer to take tasteful liberties with the tempo, *grazioso* when he wants the music to sound graceful and effortless or *legato* when he wants the music to be smooth and connected. Refinement in the rhythm can be indicated by words such as *marcato* (accented, emphasized), *pesante* (heavy, a sense of ponderousness), or *sforzando* (a sudden and marked emphasis on the note).
5. **Change in tone color and / or voicing.** This is a technique vitally important in maintaining a sense of freshness throughout the composition so that it doesn't grow stale. By altering the mix of instruments used to state the thematic material or even simply by using a first or second inversion of important chords in the progression, the composer can give a whole new life to the same melody.

Higher Level Psychological Analysis[4]

At a much higher level of analysis, and based upon a great deal of scientific literature, the single most important underlying psychological foundation of musical composition is **the creation of tension**, followed by **the release or resolution** of that tension. The listener experiences an emotional response with the release of tension. Very closely related to this concept is that within the tension is also an **expectation that the release will happen.** Another way of saying this is that music deliberatively **sets up an expectation** by creating tension and then ultimately **fulfills that expectation** by releasing or resolving the tension. Of course, composers often set up the tension, the expectation and then **delay** the release / fulfillment, sometimes for lengthy periods of time. This, too, is a powerful tool for creating emotion because a delay in our fulfillment can actually be quite delightful.[5]

Tension / Resolution → Emotion

Expectation / Fulfillment → Emotion

Repetition / Contrast → Emotion (and prevents boredom)

Because of our evolutionary heritage, we humans can **anticipate.** We are very good at this and we are very good at perceiving **change.** We predict the future, using the large frontal lobes of our brains. And we take delight when our predictions turn out to be correct, even if we've predicted a negative event. At least we can thereby prepare ourselves for the event, the change, because we were smart enough "to game" the future accurately. Certain other animals, such as apes and ravens, also display this ability and can plan ahead to anticipate their needs based upon past experience. But humans have developed this skill far more than any other animal, including our closest living relatives, the chimpanzee and bonobo.

A good composer takes full advantage of this human ability to anticipate the fulfillment of an expectation, as well as the psychological discomfort of musical tension from which we seek relief. A prime example of this phenomenon was described in Chapter 6 when we discussed harmony and chord progressions. You may want to review parts of that chapter. A highly illustrative example of tension / release is the progression from the V^7 to the I chord. In C-major, this would be the progression from G^7 to C-Major. The voicing might be G+B+D+F progressing to G+C+E+G'. The G^7 chord is filled with tension and within it resides the seed—and the **expectation**—for its resolution (the release) to C-Major. G^7 is therefore a chord that **leads** naturally to C-Major.

While the V^7 to I progression is an obvious example, and a progression used repeatedly throughout all of Western music, this technique of setting an expectation followed by its fulfillment may also be far more subtle. It can easily result from the simple relationship between two notes in a melody, with one setting the expectation for the next. It may be through the use of a *rallentando*, a slowing of the tempo, thereby setting an expectation that the composition is about to end. A relatively sudden change in the key (the tonal center) of a composition can signal that the character of the piece is about to change and this expectation may be fulfilled by an entirely new chord progression or tempo. Similarly, a sudden change in dynamic, especially from *fortissimo* to *pianissimo*, or the reverse, can startle the listener, increasing awareness that the flow of the music is changing. In each of these cases, and in innumerable others, the composer sends a sonic signal to the listener that causes the listener to anticipate a change of some kind, and then the composer fulfills the anticipation, thereby creating emotion.

Musical Form

Perhaps the most frequently used expectation / fulfillment technique is another important component of composition, namely, **repetition** and **contrast**. Think back to Chapter 5 and our discussion of the simple melody: "Twinkle, Twinkle, Little Star." Recall the basic form of that melody: A – B – A. The first measures of the tune are exactly repeated at the end, but there is a contrasting middle section (the B- section) that adds variety, a touch of surprise, and a bit of tension. Still, because of the common forms of Western music, the listener **expects** the A-section to be heard again, and of course, it is repeated, bringing **release, resolution** and a sense of completion / balance / symmetry to the tune.

Musical form, then, is a pivotal component of the composer's toolkit. Western classical music especially emphasizes defined form as a way of organizing very long compositions of an hour or more, such as the symphony or the concerto. But form is found everywhere in music, and nearly always involves the repetition of parts of the musical thought, interspersed with a contrasting element(s) of some sort.

One aspect of musical form is the *ostinato* (from the Italian meaning *stubborn).* Composers find this technique in any of its several manifestations to be very useful in creating expectations for the musical line. It's defined as a repetitive phrase or motif, or sometimes chord progression or rhythmic configuration, usually underlying the melodic line. For example, in *Bolero,* Ravel uses a constant, unwavering, triple meter base figure to undergird the melody as played by various solo instruments and combinations. This unwavering, "stubborn," persistent character tends to drive the music forward. An *ostinato* is frequently used in popular music as well, providing propulsion, and also a "foundation" as our brains process the music. There, it's sometimes called a *riff.* The *ostinato* becomes something that the listener relies upon sonically.

The Hook

In some cases, simple repetition of a melodic element is by far the predominant organizing emotional center of the musical piece. Many contemporary popular songs are like that. Everything in these songs revolves around something that composers call "the hook." The hook is either an easily remembered musical phrase (melody and harmony) or a lyric, or most often music and lyric combined. Some might call it "catchy" because we catch it almost like an infectious agent. For that reason, the composer's ideal hook becomes an "earworm," a musical phrase that we just can't stop hearing in our mind's ear and singing out loud, no matter how hard we try. This can actually become annoying when it persists. Good hooks often earn composers and performers many millions of dollars. Composers know that the components of the hook usually include a driving, probably somewhat syncopated rhythm, a simple easily sung melody, and a lyric that's not very complex.

Let's cite some examples. What do you hear in your mind's ear when you simply read the following lyrics?

Stop! In the Name of Love! (The Supremes)
A-B-C! It's as Simple as A-B-C! (The Jackson 5)
Y-M-C-A. (The Village People)
I Want to Hold Your Hand (The Beetles)
Dancing Queen! (ABBA)
R-E-S-P-E-C-T (Aretha Franklin)

Undoubtedly, you "hear" the hook that made these songs into hits appreciated by billions of people globally. Of course, not all hooks are nearly as catchy as these, but most contemporary popular music employs some kind of hook, even if it's basically only the refrain (the repeated ending section) of the song.

Western classical music sometimes employs something like a hook, though the composer may or may not have been aiming for exactly the same effect. It's usually a repeating motif, a short sequence of notes that can serve to define that piece of classical music for listeners. Examples include the opening motif of Beethoven's *Symphony No. 5 in C-minor:* G-G-G-E♭; F-F-F-D or dah-dah-dah-DAH. This motif is so catchy that it's been turned into a disco song hook.

Another example of a hook-like motif in classical music is found in the famous opening of Mozart's *Symphony No. 40 in G-minor*. Here we find a very simple motif: E♭ - D - D; E♭ - D – D; E♭ - D - D - B♭. After clearly stating the entire theme in what is classically called the **exposition,** Mozart's genius, seemingly effortlessly, improvises around this catchy little idea, transposing it into somewhat distant keys, altering its rhythm, etc. until he fulfills our expectation and relieves our tension by bringing us home again. This "coming home again" is frequently called the **recapitulation**, whereas the middle section where Mozart is "playing and exploring" intellectually with his little musical idea is called the **development**.

Arranging Music / Orchestration

Up to this point, we've emphasized the process of composition. In some cases, composition alone is entirely sufficient because the piece will be performed by the composer accompanying herself on the piano or guitar, for example. But in many other cases, completing the basic composition, consisting of melody, chord progression, and rhythm, is only about midway through the process of obtaining finished art. What remains is **arranging** or if the arranging is for large ensemble, it's referred to as **orchestration**.

Many composers do their own arranging. They assign melody, harmony, and rhythm to specific instruments or combinations of instruments to obtain the final sound that they want. They understand the sonic capabilities and limitations of each instrument of the orchestra (see Chapter 7), how those instruments sound in combination, how best to voice a chord among all of those instruments, how to achieve the proper dynamic balance between the various families of instruments, how to use the timbres of all of those instruments to achieve a desired, interesting, and beautiful tone color for the composition, the styles of playing that are possible with various instruments, the expressive qualities of each instrument, the comic potential of each instrument, the conventions used in scoring for combinations of instruments, and the list goes on and on.

To some extent, all of this skill can be taught and there are several fine textbooks[5] on orchestration and arranging for various types of ensembles, including symphony orchestras, theatre orchestras, concert bands, chamber music combinations, etc. But effective musical arranging is primarily learned by trial and error—a lot of error. Today, it's much easier than in the past because composers and arrangers have computers that can "play back" the score that they've entered and they can thereby hear a reasonable facsimile of the sound that would be produced by live musicians.

Arranging is an art in its own right. Effective arranging is difficult and even some of the great composers have produced arrangements of their own work that were not as effective as they might have been. A good conductor or other interpreter may well "correct" some of these errors to produce a more refined result. [Aside: I once saw a video by Leonard Bernstein explaining how he had changed a Beethoven orchestration to better complement the music. After all, Beethoven was deaf. He had an excuse.]

Film Music

In today's world, many composers are employed in writing scores for movies, TV shows, and video games. Interestingly, while major symphony orchestras sometimes have considerable difficulty attracting large audiences for contemporary, so-called "serious" classical music, which often emphasizes chromaticism, minimalism, and atonality (it lacks a strong tonality or key center), orchestras have little difficulty attracting crowds when they program film music in various adaptations, including the performance of the film score synchronously with the original movie projected behind and above the orchestra. This indicates that film scores have definite appeal to contemporary audiences who genuinely appreciate this fairly complex, high information content, evocative music.

Some composers have achieved considerable fame and fortune for their film work, including names such as Ennio Morricone, Alfred Newman, Bernard Hermann, Michel Legrand, John Barry, Rachel Portman, Patrick Doyle, Miklós Rózsa, James Horner, Dimitri Tiomkin, Victor Young, Max Steiner, Deborah Lurie, Aaron Copland, Nino Rota, Henry Mancini, Maurice Jarre, Elmer Bernstein, John Williams and a long list of others. Some have written iconic melodies or motifs that are not only integral to the original film but have actually become part of the fabric of our culture. Examples include the shark motif from *Jaws,* "Tara's Theme" from *Gone with the Wind,* and the motif from the shower scene for *Psycho.*

While the primary purpose of film music is to enhance the narrative of the film, and especially to enhance its emotional impact on the audience, the best examples of film music—usually when arranged specifically for concert performance—also stand on their own as memorable art. In that respect, film music is one of the most important contemporary musical genres. The power of music to create emotion in the listener is well demonstrated in this genre, at least as powerfully as we find in music from musical theatre or classical opera. For this reason, the analysis of film music provides a revealing window into the tools and techniques used by composers to create emotion.

While various film composers practice their craft in a variety of ways, their work is always tied inextricably to the wishes and vision of the director of the film. Composer and director dialogue together extensively and exchange ideas. Ultimately, however, the final decisions belong to the director. Directors have been known to fire a composer, even after an entire score has been completely written, to replace him or her with another composer. Other directors tend to employ the same composer repeatedly for almost all of their films. Think about the teams of John Williams and Steven Spielberg or Bernard Herrmann and Alfred Hitchcock.

What, then, are some of the purposes, devices, and techniques of film scoring? Here is a small sampling for you to consider as you watch your next movie:

1. Film composers will often create a motif for each character in the film. When the character is about to appear in the next scene, the audience will hear the character's

theme toward the end of the current scene. That's a powerful cue that sets an expectation, which is then fulfilled. It also acts as a "bridge" from one scene to the next, guiding the film goer and maintaining interest.

2. Film composers frequently use "underscoring" beneath dialogue to reinforce the emotional content of the dialogue, whether that content is poignant, dramatic, frightening, ominous, romantic, physically energetic, violent, or comic. Even without dialogue, music is often used to undergird and to add power to drama. Think of the very famous and shocking motif from the shower scene of *Psycho.* Interestingly, director Alfred Hitchcock wanted no music at all for that scene but composer Bernard Herrmann disagreed. Fortunately, Herrmann won Hitchcock over, simply by showing Hitchcock how much power that his iconic motif added to the scene.
3. Film composers will often use music to draw attention to particular parts of a scene. For example, ominous music may signal to the audience to "watch out" for what is about to happen in the narrative, even though the current scene at that point seems benign.
4. Film composers often employ expressive, beautiful melodies that run throughout the entire film to create an emotional context. Examples here include "Tara's Theme" from *Gone with the Wind,* the main theme from *Lawrence of Arabia*, the title song from *The Way We Were*, the title theme from *Love Story*, the main theme from *Out of* Africa, the main theme from *Somewhere in Time*, and the main theme from *Titanic.* Frankly, such a prominent theme may be the best part of the entire film, though the director of course would rarely admit it. By the way, the quality of this theme often determines whether the composer will receive an Academy Award (Oscar) nomination.
5. Film composers will sometimes use a particular popular song as a dramatic lynchpin. A wonderful example of this technique is from the best film ever made, *Casablanca* (in my humble opinion). It's rumored that the film composer, Max Steiner, disliked the song "As Time Goes By," which was written for a 1931 musical by Herman Hupfeld, even though the director, Michael Curtiz, wanted to use it prominently and dramatically. In the end, Steiner achieved a beautiful result and wrote a score that worked perfectly for the film.

Let's take a specific example of film music written by one of the greatest film composers of our time, John Williams. In the *Star Wars* series of films, directed by George Lucas, "The Force" is a prominent "character" throughout all of the films, though it is never personified, never played by an actor, nor does it speak a single word of dialogue. Some have interpreted The Force as a deity, others as cosmic energy, and others simply as a convenient dramatic device (a conceit) to organize the narrative of these films. Regardless, Williams wrote what seems to be the perfect, simple melody that appears in the score when The Force is mentioned or where its effect is important to the narrative. The audience immediately, if not consciously, thinks: "Ah, The Force!" This melodic figure illustrates much of what we've been discussing here about film music and about music in general.

The usual key for "The Force" theme is G-minor, using primarily the **ascending** melodic form of the G-minor scale. Recall that the ascending melodic minor scale (as opposed to the natural and harmonic forms) raises the 6th and 7th notes by one half step from the natural minor but returns to the natural minor when the scale is **descending**. As we shall see, climactic notes in the theme are harmonized with **major** chords. This combination of a minor key center and major chords creates exactly the tension that Williams desires, in which The Force is defined as possessing a negative or "dark" side (G-minor) and a positive, hopeful or "light" side (major chords).

You can listen to "The Force" theme in many places online to obtain the rhythm, which is slightly martial in character, but if we analyze the theme, it has just four parts:

The first statement: D – G – A - B♭ - C - B♭ - D

The second statement: D – G – A - B♭ - D - B♭ - G - D' - C'

The third statement: D – G – A - B♭ - G - D' - B♭ - G' - G

The fourth statement: B♭ - A – G - D' - B♭ - G – D – D – D - G

In terms of harmony,

The first statement: entirely G-minor tonic, the home key, the "dark" key;

The second statement: G-minor until the final C', which is harmonized with a **C-Major** chord (the IV chord of the ascending melodic G-minor scale), immediately followed by a **D-Major** chord (second inversion), which is the V chord of the ascending melodic G-minor scale, and acts as a superb leading chord to

The third statement: harmonized again with the "dark" G-minor until the climax of the theme at its highest note, G', which is harmonized with the surprisingly appropriate, **E-flat Major**, or the VI chord in the descending melodic (natural) G-minor scale, leading into

> The fourth statement: G-minor followed again by the leading **D-Major** taking us back again to the tonic, G-minor, resolving the tension and lending a sense of completion to the theme.

This simple theme appears repeatedly throughout the *Star Wars* films, with different orchestrations, instrumental timbres, harmonic voicings, and tempi but always with the same basic harmonic structure, constantly reminding the film goer that The Force has both a "light" side and a "dark" side. It is a fine example of the craft of the film composer. With a relatively simple, immediately accessible musical device, Williams has brilliantly devised a theme with harmonic underpinning to **define** an unusual film "character" that never appears on screen, is never played by an actor, and has no voice at all, other than its theme. This is what film music, at its best, can accomplish.

Science and Musical Composition

As both a practicing scientist and a professionally performing musician who has done some composition and arranging, I will make a bold and perhaps controversial assertion: composing music has much in common with the way in which a scientist examines Nature, especially in the case of applied science, which seeks to find useful applications for scientific findings. My evidence?

1. A composer begins with some education in the basic characteristics and properties of sound, much experience with the emotional impact of various melodies, harmonic progressions, and rhythms, as well as a set of "methods" and "rules" often learned from others that enable her to compose. Scientists begin with some basic training in the properties of Nature, a set of analytical and intellectual methods to examine Nature, and a set of standards for assessing verifiable evidence and in turn the plausibility of ideas.
2. A composer develops "ideas" or "notions" about what might work to achieve a particular emotional effect in the listener. A scientist develops "ideas" or "notions" about Nature and its properties.
3. A composer often uses a trial and error approach, acquiring evidence, to determine whether her new musical idea achieves the desired end result. A scientist also tests ideas about Nature, acquiring verifiable evidence, often by trial and error.
4. Ultimately, a composer creates a work of art with the emotional impact upon the listener that she desires. Ultimately, the scientist creates an explanation about Nature consistent with all known verifiable evidence and in the case of applied science, uses such explanation to make a useful invention.

As we've observed repeatedly in this book, from Pythagoras onward, scientists and musicians have been in constant dialogue throughout history. The problems and mysteries of music have frequently stimulated scientists and mathematicians to inquire about the physics of sound, to examine the fundamental harmonies of the Universe, to study the psychology of perception and emotion, to study anthropology and ethology as related to music, as well as to investigate many aspects of neurological function. In return, scientists and engineers have given music and musicians important and useful information about the physics and mathematics of music, as well as innumerable practical tools and devices, ranging from much improved and more reliable musical instruments, to acoustically effective concert halls, to a more effective use of the psychology and neurobiology of music, to superb high fidelity recording and sound reproduction technologies, and as mentioned above, to computers and software that make composition and arranging much easier and faster.

We are nearing the end of our inquiry about music and science, but some of the most intriguing findings are yet to come. Next, we'll discuss the evolution of musicality in our species, in particular we'll ask whether we are the only musical species on our planet.

CHAPTER 13

Our Musical Humanity

"The true beauty of music is that it connects people."
Ron Ayers
Composer, Vibraphonist

A handsome, four-month old male Double Yellow-headed Amazon parrot (hatched in captivity) joined our family about 15 years ago. Partly for the alliterative effect, we named him Perry (the parrot). This highly intelligent bird was so young that for two months we fed him warm gruel twice per day with a syringe before we could finally wean him to solid food. As a result, Perry became tightly bonded to the household and immediately began to mimic many sounds that we made. Double Yellow-heads are one of the three most proficient parrot species in the world in mimicking sounds that they hear in their environment. The other two species are African Grey parrots and Blue Front Amazon parrots.

The first thing that I taught Perry was the tune for "Yankee Doodle." He quickly learned to repeat my whistle of the first eight bars of this familiar, simple melody with perfect pitch and accurate rhythm. That tune is still our "call" to each other. With time, Perry acquired numerous other sounds, sometimes by being deliberatively taught through simple repetition, but often apparently spontaneously on his own. For example, he noticed that when a wet cleaning cloth was rubbed on our granite kitchen countertop, it made a squeaky sound, which he quickly began to mimic precisely. Now, when he sees a cleaning cloth rubbed on any flat surface, we hear the squeaky-squeaky sound coming from his cage.

His vocabulary seems limitless—he's constantly learning new words and sometimes long phrases. Perhaps most surprising to me, he vocalizes in the context of the situation. For example, when he observes visual or auditory cues that we are about to depart the house, he vocalizes: "Got to go bye-bye. Got to go bye-bye." And then for good measure, he adds the beeping sound made by the household's alarm keypad as the code is punched in. When he wants to rest for the evening, he will vocalize: "Time to go night-night. Time to go night-night."

Perry will also improvise sounds. For example, with the "Yankee Doodle" motif, he will change it slightly, sometimes "whistling" the entire phrase, sometimes not, and sometimes adding what appears to be his own variation.

Perry finds it easier to mimic the voices of women, probably because of the higher pitch, and he has a wide variety of "Hello's," each with a recognizable inflection and accent that he's learned from various women whom he's met. Quite humorously, Perry actually deceived the husband of a friend of ours, convincing him that his wife—who had a distinctive southern accent—was in the house, when in fact, she was not. All of this from an animal with a brain the size of a walnut in a one-pound body.

Complex Vocal Learning[1]

Scientists refer to this behavior as **complex vocal learning**, abbreviated hereafter as **CVL**. The ability is not at all restricted to birds, though of course they are most often identified with mimicry. Several groups of mammals also exhibit CVL. These include the cetaceans (whales, dolphins), the elephants, and bats. The most well-known example is probably the so-called song of the male humpback whale, much investigated by ethologists (scientists who study the behavior of animals). As very social animals, different groups of humpback whales "sing" different song patterns and the vocalizations change over time. These songs apparently serve as a mechanism for bonding the pod (the family group) of whales together.

The scientific literature presents examples of CVL in hummingbirds (which seem to have individual calls for each of their offspring), songbirds of various kinds, as well as mammals. Nearly half of all birds are songbirds and many species apparently learn their specific songs from a mentor, have recognizable variations of their songs when they are calling to specific offspring, apparently have a "song syntax," and tend to favor the consonant fourth and fifth intervals that humans also favor.

Recall that the progenitors of birds diverged about 300 million years ago, separating from the line of reptiles that eventually gave rise to mammals. Therefore, we are 300 million years of evolution distant from birds, which is quite long in evolutionary time. The architecture of avian brains differs significantly from the mammalian form. Yet, at least one gene, the so-called FoxP2 transcription factor, is apparently involved in the vocalization of both birds and humans.[2]

Of course, many sorts of animals commonly communicate using sound. Amphibians, reptiles, birds, mammals and even insects often use various sounds to signal to each other, especially regarding the presence of predators or prey, as well as to communicate sexual interest.

CVL is different from these fixed, probably instinctive vocalizations. With CVL, ethologists observe an ability of the animal to **alter** its vocalization in an apparently non-instinctive, flexible way. It's this element of adaptation to the situation, or change over time, that distinguishes CVL from the more usually observed animal vocalization.

If music is "sound organized against time," it would appear that CVL might represent a form of animal musicality. Of course, in addition to changes in pitch, the other part of our working definition of music (Chapters 2 and 3) is time. Are animals able to entrain to a pulse as most humans can so easily and spontaneously?

The answer is a qualified yes. One of the famous and most notable examples of apparent entrainment to a pulse is found in a male cockatoo named Snowball, whose antics are the subject of numerous videos available online. Careful examination of these videos, however, reveals that the "dancing" to a beat is imperfect and not as synchronous as we would observe with humans. Moreover, Snowball seems to be limited in the range of beats per minute to which he can entrain. He appears to be most accurate at about 120 bpm, whereas even non-musician humans generally can entrain flexibly over a range extending from 40 to about 180 bpm.[3]

A recent report documents the ability of a sea lion, which has not been shown to display CVL in the sense that we've been discussing here, to learn to entrain to a beat. Moreover, the entrainment is not simply in response to the heard pulse, but rather occurs even if a beat is missed in the musical source. This finding indicates that CVL and entrainment to a pulse may be independent phenomena in animals, though some scientists argue that the two are inextricably connected.

So how about our closest relatives, other primates? Because humans are musical, we might think that our closest relatives evolutionarily may also be musical, even if at a very primitive level by human standards.

This is an area of science with much disagreement among primatologists and ethologists and the data are often contradictory, confusing, and difficult to interpret. The bottom line, frankly, is that we just don't know.

First, there is little verifiable evidence that any primate species, besides the human, is capable of demonstrable CVL. All primates communicate with sound, of course, but the sounds made by these animals do not display the characteristics of flexibility that we observe with CVL in the other species described above.

Secondly, primates other than humans seem largely incapable of entraining accurately to a pulse. They can be taught to clap their hands, for example, to mimic a human who is clapping at a specific tempo. But they use visual cues to decide when to bring their hands together and are always "behind the beat," unable to continue if the human stops clapping.

On the other hand, some studies have shown that chimpanzees prefer certain kinds of human music over silence—but only certain kinds of music. In the best study to examine this question, scientists found that chimps preferred African and Indian music over silence, but preferred silence over Western or Japanese music.[4] Of course, we can't know how chimps perceive music or how they process the sound.

With regard to keeping musical time, one report describes a chimp named Barney drumming spontaneously on a barrel with an even tempo. But that's a single report. Perhaps, the great confusion in studying musicality in non-human primates is that the ability is highly variable and uncommon, which would of course produce inconsistent data.

Evolution of Musicality in Humans[5,6]

All of that still leaves us with the question of musicality in humans. Of all species, we are apparently by far the most musical, and along with language, musicality is found universally in every human group. Could both our use of language and our musicality be traits that **define** our humanity and distinguish us from all other species? Could our musicality and our language be functionally related? And if so, which came first? Language or music? While the advantage of language fluency seems obvious, what is the reproductive or survival advantage of being musical—if indeed such an advantage exists?

One piece of verifiable evidence is clear. All human cultures ever examined are musical. What is also clear is that music takes on an immense variety of forms, even in today's world, and probably was even more diverse for early humans, when populations were much more isolated from one another. In this book, we've focused on modern Western music, arising in Europe in the 17th century, and evolving into many different subtypes over the last four centuries. This music, based upon the Greek-invented diatonic scale, as a subset of a 12-tone equally tempered division of the octave, has penetrated throughout the globe and has influenced the music of many different cultures.

Yet, there are peoples completely unfamiliar with Western music.[7] Such cultures are becoming more and more difficult to find but one example was reported recently in the prestigious scientific journal *Nature*. The Tsimane' are a tribe living deep in the Bolivian Amazon, isolated from Western music. Like all groups of humans, they do have music, but with an uncommon twist. The Tsimane' make flutes and enjoy playing sequences of pitches. But they do not play pitches simultaneously. Therefore, the Western conception of harmony is foreign and completely unknown to them. When the anthropologists who wrote this report played tones simultaneously, the members of the tribe claimed no preference akin to Western preferences for consonance or dissonance and did not prefer perfect fourths or perfect fifths, indicating that such preference may have a cultural component. Of course, because musicality is influenced by both nature and nurture, it could be that there are genetic (biological) as well as cultural propensities for harmonic preferences in this isolated population.

One thing that all of the world's forms of music-making do seem to have in common are the physical acoustical principles of musical instruments, which we reviewed in detail for Western musical instruments in Chapter 7. There we described the four families of instruments: strings, woodwinds, brass, and percussion. All of the world's musical instruments can be assigned to one of these families based on the mechanism by which vibration is achieved. String instruments of other cultures often have resonators shaped very differently from Western instruments. The woodwinds may be differently shaped but they still employ either an air or cane reed. Indeed, like the use of the air reed, the use of cane as a vibrating material appears to be quite ancient. "Brasses" may well not have valves and may well be made from other metals or materials but buzzing lips still act as the sound generator. And, of course, percussion instruments come in a plethora of sizes, shapes, configurations, and types, but nonetheless produce sound when struck, stroked, or shaken. Arguably, although we cannot be certain, percussion instruments were the most ancient of all musical instruments, with the exception of the human voice, of course.

When ethnomusicologists study music from various cultures, they begin by describing the scales used in the music of that culture. This is the most recognizable and salient difference between the music of the world's peoples. And it's a great place to begin to show the diversity of the world's music.

Of course, the physics of sound and the biology of our perception provide the constraint. The concept of the octave is universal and the concept of dividing the octave into intervals also seems universal. The differences reside in the number of intervals and the distances (in terms of frequency) between them, especially whether the intervals are evenly or unevenly spaced between the octaves. Please understand that musical cultures tend to use a variety of scales for different purposes and it's usually a mistake to assign just one scale to all of the music from a given culture. Let's make a quick summary of a sampling of scales from around the world to provide just a hint of the variety:

Musical System	*Type of Scale(s) Used Frequently*
European / Western	A variety of scales as described in Chapter 5; Smallest interval is half-step; Primarily based on 12-tone equal temperament; From those 12 tones, 7 are usually selected.
Middle Eastern	Octave divided into 22 unequal intervals; Smallest interval is quarter-step; From those 22 intervals, 7 are usually selected; Example: C – D-flat – E - F – G – A-flat – B – C'
Asian Indian	Octave divided into 24 unequal intervals; Smallest interval is approximately a quarter-step; Many different scales are derived from these 24; Usually 7 tones are selected for a particular piece; Many of the scales are identical to Western music; Others use skillful quarter-step ornamentation; Bollywood music and classical Indian music differ.
Chinese Music	(Classical) Primarily employs the major pentatonic scale; Does not employ equal tempered tuning.
Japanese Music (Classical)	Primarily employs the minor pentatonic scale; Example: A – B – C – E – F – A'
Sub-Saharan Africa	Wide variety of scales with 5, 6, and 7 tones Example: D – F – G – A – C – D'
Native American	Wide variety of scales, mostly variations on pentatonic; Range from 3 to 7 tones.

Ethnomusicologists and music anthropologists also study the differences in how the scales are employed, the ornamentation of the melodic line, the harmonic patterns and complexity of the music, and of course the differences in rhythms preferred in various cultures. All of these factors and others become part of the musical language of the culture. In addition, as we discussed in Chapter 6, most musical cultures outside of Western Europe tend to emphasize rhythm and melody over complex harmony. It's not that harmony does not exist but rather that it is usually far simpler and less important in the music.

Another difference between cultures is apparently the "formality" or "informality" of most of the music heard in that culture. In Western societies, part of our musical experience is to "attend a concert." Unlike most other cultures, such a concert may be produced by hundreds of musicians, all performing together. On the other hand, yet another part of the Western musical experience is the ubiquity of music. We hear it everywhere. It becomes part of the background of our lives.

Many other cultures rarely present "formal" concerts and therefore music tends to be much more intimate, with smaller ensembles of musicians and a closer connection between musician and listener—more like a jazz club in the United States, for example. This, too, is part of human musical diversity.

So, how did both musicality and the great diversity of its expression actually originate? How much about this phenomenon is well documented and how much is pure speculation?

Humans (*Homo sapiens)* diverged from our last common ancestor with other hominids about 250,000 years ago, plus or minus 50,000 years. This number is not completely certain, but it is definitely on that order of magnitude. For much of our history, there were plenty of other hominids around, including the famous Neanderthals, and we now know that the species demarcations between hominids were not very strict because early *Homo sapiens* in Europe did interbreed with *Homo neanderthalensis* and approximately 4% of our DNA (if we are of European descent) can be traced to that interbreeding. We should think of the evolution of *Homo sapiens* much more like a branching bush, with various hominids co-existing, rather than a direct line with other hominid species dying out instantly as the *Homo sapiens* line continued.

One of the great challenges in studying these ancient humans is that culture and the ability to use language are not fossilized or otherwise well-preserved. This means that the specifics of the origins of both language and music are lost in the mists of time and all that remains are clues, about which we can speculate, but can't yet reach firm conclusions. We must rely on a combination of fossils that indicate the capabilities of the brain and the state of the vocal apparatus, together with artifacts left behind by these early hominids. Still, as we learned in Chapter 1, science begins with imaginative inquiry and the generation of ideas. The more difficult task, especially in this case, is often to find the verifiable evidence to support the idea.

One wonderful clue is a 40,000-year old bone flute, made from the hollow wing bone of a griffin vulture. Found in a cave in what is now southern Germany, the flute dates to the first settlement of that area by modern humans. The flute would certainly have been challenging technically to make, indicating that music was important to paleolithic Europeans, with apparent benefits worthy of considerable effort and energy on their part.[8]

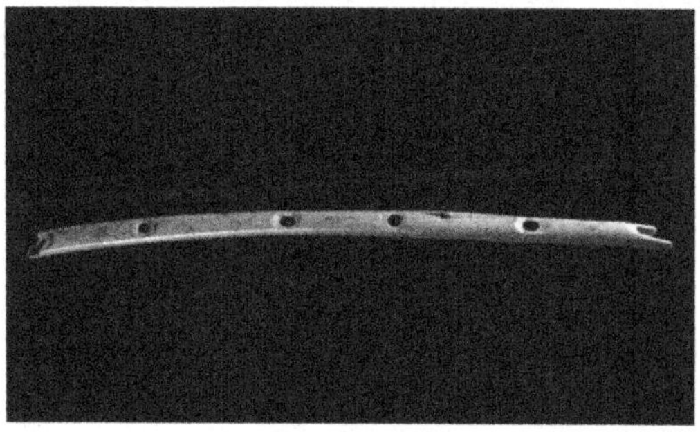

It's reasonable to think, then, that music was valuable, perhaps very valuable, to early humans. The sophistication of this flute indicates that the technology for music performance on instruments of various kinds was being developed long before that period. But what purpose did music serve?

Darwin's Theory of Evolution by Natural Selection

At this point, it may be worthwhile to review the basic principles of the theory of evolution by natural selection. This idea was elegantly articulated and defended by Charles Darwin in the mid-19th century and is now supported and enlarged by abundant verifiable evidence from many scientific disciplines, including plant and animal physiology, cell biology, biochemistry, molecular biology, molecular genetics, population genetics, embryology, comparative anatomy, ethology, anthropology, and of course paleontology. In fact, not a whit of verifiable biological evidence has ever been found that does not fit into the theory of evolution, though it is quite simple to imagine evidence that would completely undercut and destroy Darwin's brilliant idea. [Aside: When I was a graduate student at The University of Michigan, working on my doctorate in cellular and molecular biology, my fellow grad students and I would drink beer late into the night after hard work in the laboratory, sometimes imagining unusual, clever findings that would **destroy** Darwin. Needless to say, not one of us ever got a Nobel Prize for a convincing counter-argument or counter-evidence to Darwin's theory.]

Evolution by natural selection is quite simply the cornerstone, the foundation, and the scaffolding for all of contemporary biology and medicine. Its basic principles provide the intellectual underpinning for everything related to the life sciences and the benefits derived from them, including technologies for increased agricultural productivity and the discovery of life-saving medicines.[9]

Simply stated, the process of evolution can be thought of in contemporary terms with five concepts:

1. Random, constant **mutation** (change) occurs in the genetic material of all organisms. Sometimes this can even occur through the acquisition of entirely new genes from other organisms, such as viruses. The key here is that genetic makeup constantly and spontaneously changes;
2. While the vast majority of such mutations (changes) are either neutral or harmful, some are advantageous to the organism;
3. Whether a particular mutation gives rise to an advantageous as opposed to a disadvantageous phenotype (the actual physical characteristic in the organism produced by the mutation) is determined by whether it confers a greater ability for the organism to pass the mutation on to offspring. This is called reproductive advantage or greater **reproductive fitness**;
4. The process of **natural selection** simply means that mutations that enhance the reproductive fitness of the organism are **maintained** in the population, whereas those with the opposite effect are not;
5. Over millennia, sufficient numbers of mutations accumulate in a given population of organisms so that the organisms gradually **evolve into** an entirely **new species**, often remarkably different from their progenitors and / or from other contemporary species.

Human Musicality

So, where does that leave us with this trait of human musicality, found so universally throughout our species? How can we view musicality in the light of what we know about evolution?

First off, as shown clearly in Chapters 9, 10, and 11, music exists only in the neuronal firings of our brains. The architecture, cell biology, biochemistry, and molecular biology of the brain obviously depend upon genes and it follows that musicality therefore relies upon our genetic makeup in some way, at least facilitatively, if not directly. Incidentally, it's worthwhile noting here that perhaps 5% of the human population is non-musical. Such persons exhibit a trait called **amusia** or **musical anhedonia**. Sometimes this could be due to developmental error, trauma or injury. But often it seems inherited. We've all heard comments about someone being "tone deaf," unable to "carry a tune" with accurate pitch. Amusia can occasionally be far worse than that. To a person with such a form of amusia, music sounds like the clanging of pots and pans. To all persons with amusia, music makes no sense whatever, it's confusing, and certainly gives no pleasure. The trait is often accompanied by certain language deficits and is reflected in unusual neuroimaging patterns. Perhaps more interesting, the existence of amusia further indicates that musicality has a genetic component, even though we don't understand the specifics very well.[10]

Another observation about musicality is that proficiency in music is variable in the human population. On the one hand, we can observe musical geniuses and on the other, we find those who may appreciate music but know nothing about it, don't really care to learn it and certainly don't care to perform it. Traits with such a range of presentations (other examples include intelligence, physical height, weight, strength, disease resistance, sexual preference) often have both an environmental (nurture) and a genetic (nature) component. Moreover, very often, such traits are determined not by a single gene, but rather by many, perhaps hundreds of genes, all interacting together to produce the particular trait. These traits are therefore said to be polygenic. Based upon all of the verifiable evidence that we have, musicality certainly appears to be polygenic.

When Darwin himself thought about musicality, he postulated that the ability to make music would enhance the success of an early hominid to attract a mate, or many mates. For Darwin, musicality was all about sex appeal.[11] This isn't an unreasonable idea. Think of rock stars and their frequently innumerable sexual conquests. Think of young Native American males who were expected to woo potential mates with music, as well as with hunting prowess and bravery in combat. Think of high-born young ladies in Victorian England who were expected to have some proficiency with song and keyboard to demonstrate their suitability as intelligent, attractive wives worthy of the aristocracy.

But as important as attracting sex partners (leading to the act of coitus and subsequent insemination), the successful transmission of genes relies upon the nurturing of infants. Human infants have the longest period of dependency on their parents of any species. Why? Our big brains. In fact, if our fetal brains were much bigger, our heads could never pass through a reasonably sized birth canal. Therefore, we spend many years of dependency as our brains grow even larger. We are born with great neuroplasticity, which enables us to adopt the cultural norms, including the language and customs, of the group into which we are born.

So how do mothers nurture infants? First, with the biological essentials—the production of milk, the contact with their bodies, the defense from predators. But there is also the cooing, the warm comforting vocalizations, and the lullabies. Could music, and language, have originated in the mother-infant bond? Is it in these loving, comforting vocalizations that language and music began? We may never know. But it is an intriguing thought. And such a thought may mean that music **preceded** language. Could lullaby have led to language?

Ultimately, music may well have been perpetuated because of the enormous survival advantage of bonding together as a group. Synchronous, rhythmic movement combined with unison vocalization is an extraordinarily powerful means for binding even very large groups of humans together. We are clearly a social species. We don't do well alone. We need each other. But how do we bond? Common interests? Common customs? Common beliefs and myths? Common activities? Absolutely! And what could be a more coordinated, synchronous, common activity than singing and dancing together? With this idea, populations of humans that possessed the genetic makeup necessary to make music together were simply more likely to survive in a hostile environment.

One other proposed idea for the origin of musicality in humans is well worth mentioning here. Perhaps only the underlying capabilities necessary for musicality were directly selected during our evolution, including the anatomy required for speech, which arose about one million years ago. Perhaps acute hearing, the ability to extract meaning from vocalization, a concept of syntax, the ability to entrain to a pulse to assist in long distance running, highly discriminating pitch perception to assist in hunting, etc. were all selected directly for their immediate survival advantages. Only as a **result** of these innate abilities sprang music, and then only secondarily. This idea would suggest that music was an emergent, cultural invention—valuable of course, but an invention nonetheless, much like, for example, written language or religion, which have cultural value as well, and were both invented in different forms by different cultures. Such a notion would not make music any less important, but rather would indicate that musicality was **not directly** the result of natural selection.

Frankly, the verifiable evidence for none of these ideas is conclusive. The existence of some measure of musicality, especially CVL in whales and other mammals would indicate that such abilities are not secondary, but primary to survival. The apparent lack of much musicality in chimpanzees and bonobos, however, indicates that our own musicality could well have been a secondary outcome, an invention of our large brains made possible by other adaptations. If we accept this idea, we'd conclude that we simply re-purposed the neurological abilities developed through natural selection, possibly primarily for language, to be used for the production and appreciation of music. And, as an outgrowth of language and other abilities, music turned out to be a great boon for our survival. If this idea is correct, only the combination of abilities required to make and to execute the "musical invention" might be subject to natural selection.

We simply do not know the answer. Recall that I promised in Chapter 1 to draw distinctions between mature explanations that are well-supported and our nascent scientific ideas presently having insufficient verifiable evidence. While human evolution by natural election is consistent with every piece of verifiable evidence that we possess, the evolution of musicality, as a particular phenotype (if indeed it can be called a separable phenotype or trait), is much less clear. At this point, we have several good ideas, each with some level of evidence, but no single idea has emerged as convincing.

What seems inarguable, nearly intuitive, is that music does **connect** us with one another. When we are dancing together or singing in unison together at a memorial service or moving in coordinate waves to a pulse at a sports game or singing hymns in church or marching in lockstep or beating drums together around a campfire or loudly vocalizing our school's "fight song" or humming "Kumbaya" at summer camp—we are connected. We are bonded. We become one group, with common purpose, supporting one another. We've all had such experiences and we all know that music, especially when we actually produce it ourselves together as a group, provides intense pleasure, a result of the comforting sense of belonging and unity. Music simply does that. No other art does it as well.

CHAPTER 14

Our Musical Children

"I would teach children music, physics, and philosophy; but most importantly music, for the patterns of music and all the arts are the keys of learning."

Plato

Philosopher

Does Plato's statement above seem opinionated and even presumptuous? By the end of this chapter, I hope to show you, with verifiable evidence, that Plato may well have been highly insightful and prescient with his recommendation. I also hope to demonstrate, especially if you are a parent or grandparent, a teacher, a caretaker, or anyone else deeply concerned about the welfare of our children—and every adult ought to be so concerned—that this chapter offers information helpful to you as you seek ways to promote their intellectual development.

It may strike you as a bit odd that we've been dealing so much with science and yet a chapter about musicality in children begins with Plato's statement advocating music education above all else. Today our school systems are urged to emphasize Science + Technology + Engineering + Mathematics or the so-called STEM courses. In part, that curriculum is designed to prepare students for employment in an increasingly technologically complex world. As a scientist, my personal view is that the most important aim of a STEM curriculum ought to be to help students to think critically. The actual "facts" in learning the STEM information are less important than knowing how to ask questions, how to abandon so-called "authoritative" pronouncements from anyone or any source unless they are well-supported by verifiable evidence, how to gather excellent verifiable evidence, how to be skeptical constantly, how to understand that all evidence is provisional, and finally, how to appreciate the superlative value of powerful, robust, broad, coherent, falsifiable, and predictive explanations.

While such knowledge has been and will continue to be vital for the human race, there are other important things in life as well. The experience of beauty, giving us profound insight into our humanity, can come to us in various ways, only one of which is through an appreciation of Nature, her magnificence, and her laws. The arts, in all of their various manifestations, are clearly important as well. Our school curricula ought to be about STEAM, that is, Science

+ Technology + Engineering + Arts + Mathematics. Note: the arts would include literature, languages, the visual, dramatic, and kinesthetic arts, philosophy, history, and other elements of an encompassing "liberal arts" education.

In this chapter, I will provide some verifiable evidence that the study of music and the making of music by children at an early age are beneficial to overall intellectual development, enabling the easier acquisition of a variety of skills. But I also want to emphasize a key point. Music, and the other arts, are immensely valuable **in their own right**, and not merely because they enable the acquisition of various ancillary, albeit highly valuable, life skills. To view the arts as merely a utilitarian vehicle to achieve other ends is an insult to the artistic experience. The experience of beauty, however presented, is a vital part of being human. Indeed, I would be so bold as to assert philosophically, not scientifically, that without beauty, our lives can never be fully human. And I would argue that scientists seek beauty just as fervently as do artists.

Reviewing the Science of Our Musical Brain

As a prelude to our discussion of early music education, let's review some of the evidence that we've already documented in this book. Points relevant to the present discussion and also supported by strong verifiable evidence are as follows:

1. Our brains exhibit a high degree of plasticity–neuronal circuits can be formed, can change, or disappear–even the gross anatomical structures of the brain can change; (Chapters 9 and 10)
2. These changes occur when we place specific demands or requirements on our brains, especially with the acquisition of new skills or abilities; (Chapters 9, 10, and 11)
3. Such changes can occur throughout life but are particularly available to the young, when there appear to be "time windows" or "developmental stages" for the easiest acquisition of a skill; (Chapters 9 and 10)
4. Certain areas of the brain devoted to language and music overlap and are shared, whereas some areas are distinct; (Chapters 9, 10, 11)
5. Areas usually devoted to music processing can be re-purposed for language skills, for example, in stroke patients; (Chapter 11)
6. No human intellectual activity appears to demand more cognitive power than making music—in other words, when making music, the whole brain "lights up;" (Chapter 9)
7. Finally, we are a musical species and music-making is very likely a defining characteristic of our species, perhaps as defining as our ability to use complex language. (Chapter 13)

With so much evidence, it is easy to reason (to form the notion) that early music education for our children could be highly beneficial for their intellectual development, in particular in language fluency, because of the shared areas of the brain for the two functions. Indeed, we do know that speakers of so-called tonal languages (for example, the Asian languages, Chinese, Vietnamese, and Thai, where the pitch of the same word often determines its meaning) frequently possess something called **absolute pitch**[1], also referred to as **perfect pitch**. Absolute pitch is not common in speakers of non-tonal languages such as English (estimated 1 in 10,000 English-speaking people), where very few pitch changes are important for the communication of meaning. One exception, of course, is that in English, a rise in pitch at the end of a sentence indicates an interrogative, as opposed to a declaration.

"It will rain today," with no rise in pitch on the second syllable of "today," is taken by the listener as a simple declaration, whereas:

"It will rain today?" with a rise in pitch on the second syllable of "today" indicates that the speaker is asking the listener about the predicted weather.

Some English speakers, especially young people, have the unfortunate habit of raising the pitch of their voices at the end of many expressions that are not questions. Such habitual use of a rising ending pitch makes the speaker sound diffident and uncertain, thereby compromising her or his credibility.

The phenomenon of absolute pitch is intriguing to psychologists and neuroscientists. Those people who possess this ability are able to recognize a pitch and assign it the correct name from the musical scale. Many of them, upon hearing a complex chord, can name every pitch within the chord. With a few exceptions, the best evidence is that adults without absolute pitch are unable to acquire it in adulthood. Adults who do improve with training in precise pitch recognition may do so because they possess "latent" absolute pitch discrimination resulting from music education in childhood. While absolute pitch is certainly more common among musicians than in the general population, even most musicians rely on **relative pitch**, wherein once a note is heard, other notes can be generated or recognized in relationship to that given pitch. Relative pitch can quite easily be taught to musical adults through the process of **ear training**.

Evidence exists that most infants are born with the capacity for absolute pitch, just as most infants are clearly born with the capacity to recognize a palette of various colors. Young children learn to recognize colors by being told their names: red, blue, yellow, green, pink, etc. With a name for the sensory perception of the color, the two become connected in the child's brain. If young children are given the names of musical pitches, the evidence indicates that they can connect the pitch with the musical name. For example, a parent could play C on the piano, tell his child that it's a C, ask the child to turn away from the piano to eliminate visual cues, and then the parent would play the C again, asking the child for the name of the note. Over time, at least in some cases, and perhaps very often if we routinely taught children their notes as well as their colors, the child develops and retains absolute pitch for life.

The point of this discussion is to demonstrate once again that we all go through developmental stages. There are times in the life of our brain when it is able easily to acquire certain capabilities but then that period "closes off," and beyond a certain age, the skill simply cannot be acquired as easily, nor with as high degree of mastery. The acquisition of foreign language is certainly like that. It is far easier for children younger than nine to acquire a second language with flawless accent. Some very talented adults seem to be able to achieve that result, but it requires considerable effort, and they are often already bilingual or trilingual.

Effects of Early Music Education

The difficulty with investigating the impact of early music education on the intellectual capabilities of our children is that good studies are difficult logistically. Retrospective studies, wherein adult musicians are compared with non-musicians, suffer from many potentially confounding variables, such as inherited abilities, economic circumstances, and types of opportunities that musicians may have had as children but which were not enjoyed by the non-musicians in the study.

In contrast, a convincing scientific study would:

1. Select a group (a cohort) of children at a well-defined, specific age, with economic privilege, gender, family background, etc. well documented;
2. Divide that cohort of children at random into at least three groups (three "arms" of the study) *viz.* a control group that receives no intervention whatever, a group that receives an intervention not related to music (for example, the children would be exposed to sports, or visual arts, or some other training that would control for the attention given to the children by adults), and finally of course, a group that would receive well-defined education in music over a sufficiently extensive period of time, probably a year or more;
3. Employ standardized / validated before and after tests of the parameters thought to be influenced by music education, such as the ability to distinguish various sounds, the ability to learn language, the ability to learn mathematical concepts, etc. as well as abilities that might well not be influenced by music education.

Such longitudinal studies (going forward in time with the study group, rather than studying a group retrospectively) are expensive, difficult to conduct well, and therefore are not often reported. Still, these constitute the "gold standard" of verifiable evidence that is so important in devising a falsifiable, predictive, convincing explanation. Without such studies, we would rightfully remain unconvinced of the effects of early music education, however hopeful we might be that those effects would be positive. Wishful thinking is an enemy of critical thought. Recall from Chapter 1: scientists must always be the first to question their own data and conclusions, in part because if they don't, their colleagues will certainly do the critique. This is an essential, self-correcting aspect of science.

What are the results from some of the best studies? In one study, children were given either training in music or instruction in natural science. Those given music training improved significantly more in auditory information processing.[2] A second study with Chinese children receiving training in music indicated better verbal memory but not visual memory compared to the control.[3] In a third study, music training during adolescence accelerated neurodevelopment compared with the control group.[4]

One of the best sets of studies comes from the laboratory of Dr. Assal Habibi at the Brain and Creativity Institute at the University of Southern California. Eighty children between the ages of six and seven were selected from an economically underprivileged area of Los Angeles and were divided into three groups at random. One group began music education, a second began a sports-training program in soccer, and the third received no special attention whatever. Tests were administered every year to measure memory, language, music and speech processing, and brain development. At the beginning of the study, no differences were found between the three groups. One of the tests used was functional EEG to track the maturation of the auditory pathway of the brain.

After two years, children with music training could more accurately identify differences between melodies. They also had stronger brain responses to differences in pitch and their abilities to encode and process sound were faster. Dr. Habibi concludes from these early results that music education speeds brain development, at least with regard to auditory abilities. Confirming these tests, Dr. Habibi and her team also documented greater thickness in parts of the temporal cortex in children given musical training. This is a measure of brain maturation. They also found differences in the pathways of the corpus callosum.[5] Recall from Chapter 9 that the brains of musicians have been found to have a larger corpus callosum than non-musicians. These two anatomical differences would be consistent with improvements in auditory processing and with greater connectivity within the brain in musically trained children.

All the study results described above are consistent with the idea that early music education (1) improves auditory information processing and (2) results in anatomical changes in the brain reflective of such improvements. Auditory processing seems to encompass aspects of language processing, including auditory discrimination of the sounds of language, as well as of musical sounds. Such conclusions are conservative and are based upon the best verifiable evidence that we possess to date. Many other claims have been made for the effectiveness of early musical education on the intellectual development of children. Such claims, however, are very often confounded by uncontrolled variables inherent to most retrospective studies, including the self-selection of those students who become dedicated to their education in music. In other words, it could merely be the case that smarter, more curious, more naturally disciplined children tend to gravitate toward the serious study of music. If that's the case, then non-longitudinal, uncontrolled studies that claim correlations between musical performance and higher IQ, better test results, or higher academic achievement would be uninterpretable. Clearly, we need more longitudinal studies on all aspects of education to discover how best to improve outcomes for every student.

Adult Music Education

It is never too late to learn to play a musical instrument or to sing in a choir. Adult brains, while not as plastic as those of children, continue to exhibit some degree of neuroplasticity even into old age. Adults who either begin or reinitiate musical performance report benefits ranging from a reduction in stress, to greater socialization, to more pleasure in living. Bottom line: adult musical training may well be helpful as we age.

Thoughts and Recommendations

While considerable data show that musicians seem to preserve cognitive function as adults better than non-musicians, the conclusion is highly provisional because of the difficulty in controlling for the myriad of potentially confounding variables. Still, many people continue to make music throughout adulthood and derive great pleasure from the activity.

What should you, as a parent or grandparent, do for your children with regard to music? Well, first off, do realize that even infants are innately musical. The heartbeats of infants in the neonatal care units of hospitals will entrain to a heartbeat broadcast from loudspeakers.

The evidence indicates that parents and other caretakers ought seriously to consider the following:

- Expose children to a wide variety of types of music;
- Sing to and with children from infancy onward;

- Take even young children to live performances of music;
- Make up musical games with children, such as having them match a pattern of "drumming" on various objects;
- Dance to music with children;
- Teach children at an early age, especially at one to five years of age, the names of musical notes, just as you teach them the names of colors;
- Encourage music education as part of both the elementary and secondary school curricula;
- Remind children that music and the other arts are about beauty and that the experience of beauty is an immense privilege;
- Provide music lessons of any kind;
- Encouragingly "make" children practice, at least for a time, not in an overbearing way, but as an act of love, just as you "make" children do their homework or tidy their room to teach discipline and the pleasure of accomplishment;
- Attend the musical performances of your children.

While nearly all very young children seem to enjoy the pleasures brought by music, not all children remain enamored with making music as they grow older. Some would rather play sports, or dance, or draw, or do science experiments, or build things. But a basic exposure to music and to music-making in the early childhood years (infancy to about nine) would seem, from everything we know, to pay many dividends throughout life, not the least of which is to provide a basis for continuing musical performance throughout adulthood.

For those children who profoundly love music and possess talent and discipline, music programs (choirs, bands, orchestras) designed for children between nine and eighteen are extremely valuable, if only for socialization, for the development of diligence, for a sense of mastery, and for learning laser-like focus on task completion. But even more important than those benefits, such very fortunate children actually create beauty on a daily basis using their own minds, hearts, and hands. In the process, they most assuredly become more fully human. And isn't that the ultimate purpose of education—and of living?

In a possibly apocryphal story, Albert Einstein was being praised for his brilliant scientific work. His humble response was that because of the inexorable forward progress of science, his own findings about Nature would have been uncovered by someone else eventually. But there would **never** be another Mozart.

In a similar way, the beauty that our musical kids make, whether performing alone or with others, belongs to them as their unique creation. It is their birthright.

In the final chapter of this book, we will discuss our Cosmos and how scientists have found that its magnificence is intimately, perhaps inextricably connected to the beauty of music in ways that may well surprise and delight you.

CHAPTER 15

Cosmic Music

"The most beautiful thing we can experience is the mysterious. It is the source of all true art and all science."

Albert Einstein
Physicist, Nobel Laureate, Musician

"If you want to know the secrets of the universe, think in terms of energy, frequency and vibration."

Nikola Tesla
Physicist, Engineer, Inventor, Futurist

"The world is made of waves. But when you look at it, you see particles."

Sean Carroll
Theoretical Physicist, Science Communicator

"At the heart of matter is music."

Brian Greene
Theoretical Physicist, String Theorist, Science Communicator

Scientists do not yet understand the fundamental nature of reality. Understanding the essence of reality, at its deepest and most profound level, is probably the single most challenging, most debated, most beguiling, and most puzzling question in all of science—one so important, even in the minds of the general public, that a film about the life of the late physicist and cosmologist Stephen Hawking was entitled simply *The Theory of Everything.* While we do not yet have a theory of everything, we do have pieces, internally consistent fragments of understanding, that make some sense to us and are highly predictive, within their own domains. At this point, however, they don't tie together. They don't form a cohesive whole or a complete picture. At least, not yet.

Emergent versus Fundamental

Most of what we do in science is to study emergent phenomena. By that, we mean phenomena that are the **consequence** or the outgrowth, perhaps the inevitable result of the very essence of reality, rather than whatever is actually that essence itself.

Let me illustrate. When a young neuroscientist, a new father, looks into the eyes of his newborn daughter, he isn't thinking about his own hundred billion neurons, their neurotransmitters, and how they are leading to the immediate sensation of love that he feels for this infant human being whom he has helped to create. No. He is basking in the emotion, the wonder, the pleasure, the power of his affection for her. It's not that he's unable to explain some of the neuronal processes in his brain that result in his feelings for her. But at that moment, such considerations are simply not important to him. He is experiencing an emergent aspect of reality.

Scientists ordinarily study Nature at various fairly discrete and defined levels. In many cases, mixing those levels may even confound the science, making the task far more difficult. For example, when a biologist studies animal behavior on the plains of Africa, she is not, in that moment, much concerned with the details of the biochemical reactions initiating the neuronal firings in the brain of the animal that ultimately cause it to seek sex or food. No. She is interested in the emergent, resultant, actual behavior of the animal. That is her focus. And she can gather meaningful verifiable evidence only by focusing at that level of observation.

We scientists, with our own personal preferences, interests and training, choose levels and aspects of reality to study, investigating subjects that may range from the entire Cosmos and its origin all the way to the physics at the core of the most fundamental nature of reality itself. All levels are important. All are valuable if we are to gain a broad understanding and to produce inclusive, comprehensive explanations.

Our Scientific Search for the Nature of Reality

For many centuries, humans ascribed "existence and its fundamental nature" to some deity or another by default. In other words, humans are inclined to employ supernatural "explanations" rather than predictive, verifiable evidence-based explanations derived through scientific naturalism, especially when posing questions about the fundamental nature of reality. This is a common tendency and an apparently universal predilection of humans, one manifested whenever we cannot find an immediate natural explanation. Think of the many, many times throughout history that we resorted to using the supernatural to explain everything from plagues, to famines, to the weather, to the political outcomes of various human actions, to the motion of the planets, to the origin of diverse life forms and on and on regarding anything that we didn't understand.

Still, modern science pushes forward inexorably, powerfully, skeptically, empirically. Scientists work with diligence and determination to find coherent explanations that are both falsifiable and predictive (Chapter 1). It's precisely like that with our inquiry into the fundamental nature of reality itself.

A Simplified Sketch of the History of Our Search for the Fundamental Nature of Reality[1]

Prior to the advent of modern scientific inquiry, philosophy was the means of inquiry available to us. One form of philosophy, so-called "natural philosophy," was the intellectual precursor of contemporary science and of scientific naturalism. A remnant of this history remains with us even today because most professional scientists in the United States hold the Doctor of Philosophy degree, the Ph.D.

Philosophy ordinarily relies upon reason, argument, opinion, conjecture, assertion, and speculation to conduct inquiry. Natural philosophy, as it developed, began to add to that list the actual **observation** of Nature to obtain clues that would enhance reasoning and speculation. As we discussed extensively in Chapter 1, Western science, when it gradually gained increasing influence during the 16th and 17th centuries, and especially with its great leap forward in the late 18th century, added verifiable evidence, skepticism, empiricism, and the rejection of authority to the task of making observations about Nature.

The Greeks[2,3]

The ancient Greek philosophers were much engaged with attempting to understand the fundamental nature of reality. In the colloquial, we'd say that this was a "big deal" for them. We've already encountered Plato in Chapter 14 and learned of the value that he ascribed to teaching music to children. Of all the Greeks, his ideas about the nature of reality may be the most famous. He developed an idea of "true forms." Plato noticed that humans were able to recognize a tree as a tree, or a chair as a chair, or a rock as a rock, even though no two trees, nor two chairs, nor two rocks were exactly the same in appearance. He speculated that the fundamental nature of "chair-ness" existed in a true form, an absolute form of chair-ness, and that this true form was the fundamental nature of reality. From this idea, Plato developed the notion that various geometrical shapes also had a "true form" that was immutable. Moreover, he thought that five perfect geometrical solids could account for everything in the entire Cosmos, thereby reducing the Cosmos to a mathematical description. Of course, as with much of philosophical speculation, this notion was fun and interesting, but without a shred of verifiable evidence to support the idea.

For our friend Pythagoras (Chapters 4, 5, 6), the analysis of music was the principal impetus and inspiration for his view of mathematics as the basic descriptor of the fundamental nature of reality. The nature of reality for Pythagoras and his followers was encoded in the language of mathematics—just as beautiful and pleasing harmonies were described and encoded by ratios of whole numbers. This theme that mathematics is at the center of understanding the nature of reality will be recapitulated often in our discussion later in this chapter. Pythagoras was the first to articulate the idea powerfully—and he used music as a pivotal part of his argument. Here again, science and music were connected inseparably.

Yet another theme of contemporary physics was foreseen in fifth century B.C.E. Greece, namely, the concept of atomism, from the Greek adjective *atomos*, meaning "uncuttable" and therefore indivisible. For the Greek philosopher Leucippus and his student, Democritus, the fundamental nature of reality was about indivisible, invisible, indestructible units, and the space ("the void") between them. Of course, these thinkers had no verifiable evidence whatever. It was an idea based upon reasoning, speculation, and conjecture. Nonetheless, as we shall soon discover, the notion of indivisible particles as a centerpiece of reality would ultimately be demonstrated clearly by 20th century physics, so much so that we call it "the standard model of elementary particles."

The Scholastics

While certain aspects of science, especially astronomy and mechanics, continued to thrive in the Arab world and in China, Europe unfortunately experienced 600 years of stagnation and desperation with the fall of the Roman Empire. Supernaturalism became the predominant intellectual force behind all perceptions of the nature of reality. In the 13th century, however, the Greek philosophers were rediscovered in Europe through translations preserved and interpreted in Arabic. Aristotle's writings became the basis for a re-examination and new formulation. Thomas Aquinas, in particular, tried to find a way to reconcile Aristotle's notions with the dominant supernaturalism of medieval Europe. The results were certainly less than satisfactory. Aristotle relied upon an arbitrary "ideal." For example, because a circle was arbitrarily declared to be ideal, then the planets must move in circles. No other option was permissible. The idea of actually observing the results of "artificial experiments" was largely rejected. Reason, conjecture, argument, speculation and assertion, combined with Aristotelian ideas and with the overriding "authority" of supernaturalism, were considered to be more than sufficient to devise an accurate understanding of the nature of reality. In that view, no verifiable evidence was required.

The Birth of Scientific Naturalism: The Scientific Revolution[4]

While scholasticism did help Europe to obtain some relief from its 600-year intellectual drought, this philosophical system was nonetheless destined for ultimate rejection. Within its philosophy were the seeds of its own destruction. A disdain for verifiable evidence, and instead a reliance upon fixed, dubious "authority," was a recipe for failure, even if the aim were to obtain merely a glimpse of the fundamental nature of reality.

During the 16th and 17th centuries, Europe underwent an intellectual step-change—actually, several of them. For over 2,000 years, the earth was thought by nearly everyone and proclaimed by church "authority," based upon its interpretation of "holy scripture," to be the center of the Universe. Nicolaus Copernicus, Johannes Kepler, and Galileo Galilei (Chapter 3) concluded that their own heliocentric model was correct. The earth was **not** the center of everything, as supernaturalism and the scholastics maintained. Moreover, using his telescope, Galileo found verifiable evidence for this conclusion, as opposed to the utterly inadequate conjecture, assertion, and "authority" relished by scholasticism. Sadly, Galileo, and too many other scientists, suffered terribly at the hands of "authority" for such boldness.

Mathematics gained great impetus and importance. Galileo employed it to quantitate the results of his experiments. For example, Aristotle had asserted, absent verifiable evidence of course, that heavier objects must fall faster than lighter objects because—well—because they were heavier—and it was obvious. Galileo did the experiment! He found that heavier and lighter objects actually fell together and their identical rate of falling was quantifiable using mathematics. Falling objects accelerated predictably as a function of the square of the time. The immense power of verifiable evidence, described mathematically, was thereby dramatically reinforced. According to Galileo: "The Universe is written in the language of mathematics."

René Descartes and Gottfried Leibniz made other essential advances in developing the language of mathematics in the 17th century, providing the tools necessary to describe Nature more and more accurately in its own language. The view of Pythagoras that the essence of Nature could be described and explained mathematically was now becoming entirely credible.

The great 17th century polymath, Isaac Newton, made seminal advances in mathematics and articulated the laws of the motion of objects under the influence of a force that he called **gravity**, reaching conclusions that were completely at odds with scholasticism. For example, he showed compellingly that the planets moved in ellipses, not circles.

Other scientists made major contributions to our understanding of the behavior of gases, heat, and all types of mechanics throughout the 18th century. At the same time, chemistry was advancing from alchemy to a much more systematic approach, especially with the work of Antoine-Laurent Lavoisier in Paris. Biology, with the microscope as a tool, was still at the stage of describing, cataloging, and dissecting but those efforts served as an important basis for later advances. William Harvey showed that the heart acted as a pump for the circulation of the blood.

By the 19th century, the world had changed greatly. From the time of the Protestant Reformation in the 16th entury, Europeans had begun seriously to doubt "authority," and then with the French Revolution and the Age of Enlightenment from the mid 17th to the late 18th centuries, along with the American Revolution, the power of church and king, and their presumed authority, had declined precipitously. With the advent of the 19th century, scientists working in Europe and North America began to experience the benefits of the overthrow of both forms of tyranny, the supernaturalist and the dictatorial. Natural philosophy had transformed into real science, that is, into scientific naturalism, with its firm reliance upon verifiable evidence, empiricism, skepticism, and the rejection of authority. Indeed, scientific naturalism made only **one** assumption: "Nature does not aim to deceive us."

The 19th century brought advances at an unprecedented rate and in all branches of science. Some consider the 19th century to be the age of electromagnetism in physics, along with the precise articulation of the laws of thermodynamics (heat exchange and the conservation of energy), radioactivity, and many aspects of mechanics, including insight into standing waves on strings and in pipes (Chapter 7, "Making Music"). Great scientists such as Michael Faraday, Hermann Von Helmholtz (Chapters 4, 6, 7), James Clerk Maxwell, Heinrich Hertz (Chapters 4, 5), Marie Curie, Max Planck and others made enormous advances in fundamental and applied physics.

Chemistry was also developing. Organic chemistry was born with the artificial synthesis of urea, a compound previously thought to exist only in living organisms. This is the chemistry of the carbon atom, a versatile element upon which all known life is based. Carbon has the ability to bond with other atoms with elegant diversity to create an immense array of compounds that possess highly variable and valuable properties. In part, this is because it can form long chains with itself or combine in closed rings. Chemists learned to manipulate carbon very effectively, creating a vast array of often new compounds that had not existed before. Recall our discussion from Chapter 12, "The Composer's Toolkit," about the important scientific work of Russian composer Alexander Borodin in organic chemistry.

As wonderful as all of this was, the greatest advance of the 19th century was Charles Darwin's theory of evolution by natural selection.[5] This idea utterly changed our perspective and thinking not only about the origin of life forms on earth but also about our basic concepts of Nature itself and about how Nature actually works—with no verifiable evidence for any involvement of the supernatural. Essentially, Darwin's ideas show that meaning and purpose arise directly from the material. Such a concept was offensive to some persons at the time and still creates controversy and upset even 150 years later in certain segments of the population, who apparently feel threatened by this immensely successful, highly predictive, robust theory. Oddly, no similar offense whatever is taken about other important, equally well-supported and pivotal scientific theories such as Newton's laws of motion or Einstein's general relativity or even the "germ" theory of infectious disease. Recall that we discussed the theory of evolution by natural selection in Chapter 1 ("What is Science?"), Chapter 8 ("The Ear"), Chapter 9 ("Our Musical Brain") and extensively in Chapter 13 ("Our Musical Humanity"), so we needn't go into as much detail here.

Darwin's theory shows clearly that Nature undergoes change **on its own,** not directed or guided by a "supernatural" or "designing" force of any kind, but rather merely because certain changes are of value so as to be preserved, and others are not. Amazingly, Darwin had this insight without a mature notion of genetics (inheritance), which was not to be discovered until the very end of the 19th and the beginning of the 20th centuries. When the science of genetics finally arrived and matured throughout the 20th century, it added even more powerful support to Darwin's theory by providing a cellular (chromosomes) and then later a molecular (genes / DNA) mechanism for his ideas about evolution. Throughout the 150 years since Darwin, not an iota of verifiable evidence has ever been found inconsistent with the theory of evolution by natural selection. Rather, we continue to accumulate massive amounts of verifiable evidence that support and affirm it.

While all of this work, starting from the 16th century, dealt with emergent phenomena, it provided a critical foundation for the technological advances given to us by science, ranging from clarinets to concert halls, from electronics to engines, from medicines to metallurgy, from radios to rockets, from satellites to smartphones—and everything in between. But we were still very far from genuine insight into the fundamental nature of reality itself. In the 20th century, several geniuses gave us the beginnings and at least the hope of such an understanding. But they also gave us contradictions and controversy.

Twentieth Century Physics and Cosmology[6]

Twentieth century science begins with Albert Einstein. Even *Time* magazine (December 31, 1999) named Einstein the "Person of the Century," in part because his ideas changed our entire perception of the nature of reality and got us closer to its essence than we'd ever been before.

Time **Magazine Declares Einstein Person of the Century! December 31, 1999**

Surprising to many, Einstein's Nobel Prize (1921) was not awarded for the famous theory of general relativity that we discussed in Chapter 2 ("What is Music"). It was for something called the photoelectric effect, which we'll discuss momentarily.

His first important discovery (1905) dealt with neither of those ideas but instead provided experimental evidence and mathematical predictability for the existence of atoms. In brief, Einstein developed the necessary mathematics and the physics to make predictions about the behavior of atoms and he did this with enormous precision by studying the well-known phenomenon of diffusion and something called Brownian motion. Many scientists of the 19th century, though not all, thought that sufficient evidence already existed to make the idea of atoms highly plausible but Einstein convinced everyone. The Greek atomists had been vindicated. We were one step closer to the nature of reality.

As we discussed before, Einstein is probably best known for both his theory of special relativity (1905) and his immensely important theory of general relativity (1915). He was amazingly good at something that we've already done ourselves in this book: the *Gedankenexperiment*. Such thought experiments constituted the **imagination** part of his inquiry. For example, he imagined himself on a train that was moving faster and faster until it approached the speed of light. Then, he began to think about time and how it might "pass" for him compared with its passage at the train station of origin, relying on his knowledge that the speed of light in a vacuum is constant and finite.

Einstein was a fine mathematician, but he often sought assistance with his mathematics from those more expert than he. This was merely one aspect of his humility. In any case, he expressed his imaginative speculations about physics in the language of mathematics. Math provides great power for inquiry because it employs a symbolic language with demonstrated theorems (mathematical relationships). By working in mathematics, rather than in ordinary verbal language, physicists can thereby make **quantitative predictions** about the physical world, and when these predictions are tested empirically, physicists learn things that they might not have even considered without mathematics leading the way.

Note carefully, however. As useful as mathematical relationships are for guiding inquiry, they do **not** constitute verifiable evidence. Empiricism and skepticism reign supreme in science. Without testability (falsifiability), it's just not science, no matter how seductively beautiful the mathematics may be. In science, we must **observe** Nature, often in very clever ways, and the observation is essential for us to reach falsifiable and predictive explanations.

Einstein wrestled with certain contradictions in Newtonian ideas about motion. First, Einstein knew that, according to Newton, the perception of motion depends on the reference position of the observer and also that different observers could well see the same event but make an entirely different conclusion about how objects were moving. In other words, motion and events must be defined **relative** to other objects and thereby depended on a frame of reference. Secondly, Einstein understood that the speed of light through a vacuum was **finite and constant relative to everything and for all observers**, and that light, of course, carried the information necessary for the observer to make conclusions about the motion of objects. The speed of light is therefore also the speed of causality. Newton's Laws imply an infinite speed of light. This was a fundamental contradiction in physics. How can the motion of everything but light be dependent on a frame of reference and what are the consequences? Einstein found a resolution that required a complete revision of our conceptions of time, space, energy, and gravity.

Theories of Special and General Relativity

While the equations of Einstein's idea of relativity are complex and somewhat difficult, his final conclusions can be summarized as follows:

1. Matter (mass) and energy are equivalent and interconvertible as described by the famous equation $E=mc^2$, where E is energy, m is mass, and c is the speed of light. Mass and energy are two manifestations of the same underlying reality.[7] This conclusion results from the idea that as an object moves faster and faster by adding more and more kinetic energy, its mass increases. If it moves at the speed of light, its mass is infinite and the energy to move it would be infinite as well. Nothing can travel faster than the speed of light and nothing with mass can travel at light speed. This also means that if the speed of light (causality) were infinite, there would be no mass in the Universe.
2. Space and time are unified, forming one physical "thing" called space-time.
3. Gravity works by the bending of space-time. Thus, the earth, as it orbits the sun, is actually moving in what would be a straight line, but because the mass of the sun bends the fabric of space-time, the earth's movement merely follows the warped fabric of space-time.
4. Our ordinary, Newtonian conception of time is incorrect. Time is **relative**. It "passes" at different rates at different places in the Universe. The rate of its "passage" is determined by two things: the speed at which an object is moving and the influence of gravity on the object as space-time is warped. The faster that an object is moving, the slower that time "passes." The greater the influence of gravity on the object, the more space-time is bent, and the slower time "passes." Time therefore dilates. (See Chapter 2 for additional discussion of what this means about our perception of time.)

One of the greatest dramas of science was played out in the 20th century following Einstein's publication of his theory of general relativity. While Einstein's ideas did seem to account for the small difference in the orbit of the planet Mercury compared with the predictions of Newton's laws of motion, much more verifiable evidence was required. There was only one practical way to get it.

During a complete solar eclipse, the moon covers the sun, casting its shadow on the earth. If Einstein's ideas were correct, stars "in back" of the sun would appear to change their predicted positions as observed on a photographic plate made during the eclipse because the sun's gravity would "warp" space-time, causing the light from those stars to follow the bending of space-time. The change would be very small but Einstein's math made highly specific predictions for the change in the stars' apparent positions. Again, this is the power of the mathematics. It enables scientists to make **quantitative predictions** about the nature of reality.

The international political conflicts of World War I complicated travel to places on the globe where total solar eclipses would occur, and of course, cloud cover during the eclipse would render the effort hopeless. Still, after several attempts, and much doubting in scientific circles, Einstein was shown to be correct. The verifiable evidence was beautifully consistent with the theory. Further, numerous, even more sophisticated tests of general relativity were made throughout the 20th century without a single result that deviated from Einstein's ideas.

This made Albert Einstein the most famous scientist in the world. He had changed our conceptions of time, space, energy, mass, and gravity, and had given us a major step forward in our understanding of the fundamental nature of reality.

Still, while relativity made him publicly famous, Einstein had actually already contributed something else very important in 1905, with verifiable evidence that would ultimately lead to great controversy and discoveries that would haunt Einstein until his death, even though this same work ironically earned him the Nobel Prize. His findings had to do with something called:

The Photoelectric Effect

Before discussing the photoelectric effect, let's make a point about light. In the 19th century, scientists regarded light as an electromagnetic wave, just like radio waves. Its behavior was "wavy" and entirely predictable if we thought of it as a wave propagating in space, just as the waves of sound propagate in air (Chapter 4). The casting of shadows when light is blocked by an object is an obvious example of this behavior. This was all fine and good, except for one thing.

Physicists discovered that light would "knock" electrons (particles carrying a negative charge) off metal. Different colors of light were differentially effective. Red light, no matter how intense (bright), was ineffective. On the other hand, ultraviolet light, even if dim, was highly effective. If light were simply a wave, increasing the intensity of red light should have dislodged the electrons. But it didn't.

Einstein concluded that red light, at a lower frequency of vibration, had a lower energy, whereas ultraviolet light, at a much higher frequency, possessed much higher energy. More important, while behaving as a wave, light was also **quantized.** That is, it moved through space as discrete **quanta** (units) of energy, called **photons.** Max Planck had proposed this idea several years before, but had not provided empirical evidence for it. Einstein showed that with such quantized energy it was possible to explain the photoelectric effect, and several other light-related phenomena as well. Light (and other electromagnetic radiation) could behave like a wave **or** a particle, but never both. Its behavior depended on the experimental conditions of measurement.

Coupled with Planck's ideas, as well as those of Louis de Broglie, Einstein's work ushered in a whole new area of investigation in physics, an area later dubbed **quantum mechanics,** which was intensively developed by such luminaries as Werner Heisenberg, Niels Bohr, and Erwin Schrödinger. But Einstein died in 1955 without ever accepting the apparently bizarre interpretations of quantum mechanics, in part because they were inconsistent with relativity. Quantum mechanics makes accurate predictions about the very small—the subatomic—and relativity makes accurate predictions about the large, including everything that we ordinarily observe with unaided vision. As far as we can tell, both theories are fully respectable within their own domains. No prediction made by special or general relativity has ever been incorrect, with the latest vindication being the discovery of gravitational waves in 2016. And quantum mechanics makes it possible for us to invent modern electronics including our cell phones, lasers, and numerous other technologies.

Before attempting to explain quantum mechanics—and no one can do that anyway because no one in the world really understands it—let's discuss another area of 20th and 21st century physics that is well-documented, highly explanatory, and consistent in large measure with our intuition. It's called:

The Standard Model of Elementary Particles.

Let's start our discussion with a review of the classical structure of the atom as described largely by Ernest Rutherford and later by Niels Bohr. The atom is composed of a **nucleus**, which contains particles called **protons** (with a positive charge) and **neutrons** (with no net electrical charge). Surrounding the nuclei of atoms, we find probabilistic "clouds" of **electrons.** Certain "orbits" or energy levels of these electrons are permitted, whereas others are not. The number of electrons and protons in a neutral atom are equal. Each element (hydrogen with one proton, carbon with six, oxygen with 8, copper with 29, etc.) is **defined** by the number of protons that its nucleus contains. Electrons, however, are **non-localized**, and can move easily, either to different "orbits" or sometimes away from the nucleus altogether. Indeed, electricity is composed of electrons moving. Moving electrons carry energy and information, create magnetic fields, and are extremely important to the functioning of neurons (refer back to Chapter 12, "Our Musical Brain", and Chapter 13, "Music Changes Us").

When a photon strikes an atom, it is able to change the "orbit" or energy level of an electron if the photon possesses precisely the correct amount of energy. When the electron moves back to its initial "orbit," a photon of precise energy is emitted. This explains the observation that each element has a specific "emission spectrum" and enables the identification of elements in stars. Moreover, this means that matter is quantized as well.

During the last one hundred years, physicists have gradually developed a much more detailed "standard model" of this aspect of reality. We won't and can't in a few short paragraphs analyze or even list the massive amount of verifiable evidence for this model, nor its elegant mathematical principles, but rather we'll just summarize its major characteristics.

Nature has four fundamental forces. A force is defined as something that can cause a mass to accelerate.

- The **strong nuclear force** explains how protons stick together in the nucleus of the atom despite their repulsive positive charge;
- The **weak nuclear force** explains radioactivity, matter versus antimatter, and the "spin" of particles;
- The **electromagnetic force** explains light, electricity, and magnetism and it's the force that actually holds the atom together;
- **Gravity** explains the motion of the planets and why we ourselves don't float off into outer space spontaneously, but it is a much, much weaker force, by many orders of magnitude, than the other three forces.

Physicists doubt that other forces will be discovered. These four seem to be it. On the other hand, physicists desperately desire an equation that would combine all four of these forces into a single, elegant expression *i.e.* a unified field theory. Note: a **field** is a mathematical construct that specifies the value of the force at every point in space. So far, they haven't found it. Such an explanation would become part of the theory of everything.

The other part of the standard model is an array of subatomic particles, found initially as cosmic rays but now often generated and then revealed by smashing one particle into another at very high energies in places like the famous Large Hadron Collider. Physicists study the particles generated from these collisions, categorize them, and have now formed a consistent, apparently complete particle model of reality. The powerful mathematics developed by Richard Feynman and Murray Gell-Mann, including quantum electrodynamics, was essential for us to confirm that this description is accurate. But even at this level of reality, and as small as these particles are, we have not yet reached Nature's most fundamental level.

The elementary particles for the Standard Model include the following:

- **Quarks** (6 kinds which in various combinations form protons and neutrons; 3 quarks each)
- **Gluon** (carries the strong nuclear force and holds the nucleus together)
- **Photon** (the mass-less particle that constitutes light and other forms of electromagnetic radiation)
- **Leptons**
  - **Electron** (carries negative electric charge)
  - **Tau** (a heavier electron, very short life)
  - **Muon** (a heavier electron, longer life than Tau, component of cosmic radiation)
  - **Electron neutrino** (an electron-like particle but without charge)
  - **Tau neutrino** (a tau-like particle but without charge)
  - **Muon neutrino** (a muon-like particle but without charge)
- **Bosons**
  - **Gauge Bosons (**2 kinds, namely the Z boson and the W boson, that carry the weak nuclear force)
  - **Scalar Boson (**1 kind: the Higgs boson, responsible for conferring mass)

Please note: the existence of these particles is **definitively predicted by mathematics.** But because science relies on empiricism, they had to be demonstrated. Now, we have verifiable evidence for all of them with the recent demonstration of the Higgs boson in 2012.[8]

The combination of these subatomic particles and the four forces describe physics as we know it today from everyday experience on earth. But you should remember, and this is important: **Each of these elementary particles, and individual atoms as well, can behave as a wave or as a particle.** How they behave is dependent on how we examine them. This realization is key if we are even to discuss quantum mechanics.

Quantum Mechanics

Here's the rub in our current investigation of the fundamental nature of reality. Special and general relativity are well described mathematically and successfully explain big things in the Universe. The standard model makes a lot of intuitive sense and explains the basic construction of matter and the forces (fields) that hold it together—except for gravity. But quantum mechanics, that is, the description of the behavior of subatomic particles and even atoms—while certainly entirely verifiable based upon countless empirical observations, predictions, and the technologies derived from it—is also completely **counterintuitive** and describable only with mathematics. Moreover, general relativity and quantum mechanics do not reconcile when it comes to gravity.

Let's list some of the outcomes of quantum experiments that cause so much consternation. The behavior of the electron has been used most often in these experiments, in part because the electron is so important for our highly useful electronic gadgets. Herein, it is representative of all other subatomic particles.

First, the verifiable evidence shows that the position of an electron is **probabilistic.** The wave nature of the electron specifies the **probability** of finding the electron at a point in space. Therefore, at the subatomic level, there is no predictable certainty about the exact position of the electron or of any other subatomic particle.

In fact, the location of the electron becomes fixed **only when it is observed.** Prior to being measured, the possibility of finding the electron in any particular place is probabilistic. With observation, it becomes fixed. In fact, a famous type of experiment, the double slit experiment, shows us that observing an electron, either before or after it passes through one of the two slits, causes it to lose its wave-like properties and to behave instead like a particle. Erwin Schrödinger's equation implies that upon observation, the wave function of the electron "collapses" and the electron can be identified as a particle at a single point. This is referred to as the "Copenhagen interpretation" of quantum mechanics. You may well ask: how can merely observing an electron change it from a wave to a particle? Strange, is it not? But it implies that reality does not exist without "seeing" it. This is known as the "measurement problem" in quantum mechanics.

An alternative to the Copenhagen interpretation is the "many worlds" interpretation of quantum mechanics, wherein all possible states of the electron exist simultaneously, described within a single wave function of the Universe. But once the observation is made, the two different states branch into two separate "worlds," now eternally separate, both of which were already there initially in what is called a "superposition." What this means is that **everything**, not just small particles, is quantum mechanical.

Related to the "many worlds" interpretation are experiments and equations of quantum mechanics that demonstrate yet another critical property of electrons called **entanglement.** The measurement of an electron at one point affects the characteristics of the state of **another** electron that is entangled with it and this happens instantly at any distance between them. Einstein disparagingly called this phenomenon "spooky action at a distance" and maintained that the state of the electron was instead already determined prior to observation. Experiments in the late 20th century showed definitively that Einstein was wrong about this. Detecting the "spin state" of one electron will instantly specify the state of its partner. This is perhaps the most bizarre finding of quantum physics, but it is also the most definitive. While we have mathematics that predict this strange phenomenon, the phenomenon remains completely counterintuitive.

The fourth well-documented quantum phenomenon is called **quantum tunneling.** The wave whose amplitude represents the probability of an electron being found at a particular location can be present inside a barrier that would ordinarily always block the electron. Because the probability wave moves through the barrier, there is a certain probability that the electron can **tunnel** through the barrier as if unimpeded. And it does.

All of this is simply weird. While physicists have developed many interpretations of quantum mechanics in trying to make some sense of its weirdness, no single interpretation is universally accepted. Although we have very little doubt that the quantum properties of Nature are verifiable and that our ability to predict those properties is genuine we still do not really understand how it all happens. Albert Einstein remained convinced that quantum mechanical phenomena are verifiable but that our interpretations (explanations) must be incomplete.

Regardless of our lack of a universally accepted explanation, the equations of quantum mechanics allow us to design transistors, lasers, and innumerable electronic gadgets that have transformed our world. Even photosynthesis (the ability of plants to turn carbon dioxide and water into sugar using photons as the source of energy) and bird migration seem enabled by the quantum mechanical properties of Nature. Quantum mechanics may well be involved in the perception of odor, in the metamorphosis of amphibians, and in evolution itself. We are forced to conclude that quantum phenomena are real, can be expressed with extremely accurate predictability by mathematics, and play important roles in our technology and even in our biology. Despite our intuitive protestation, this means that all matter, including we, ourselves, have the properties of waves—much like sound and therefore, much like music. Are we, as humans, actually a form of music?

Towards a Synthesis and Reconciliation: But the Mystery Yet Remains

General relativity and quantum mechanics seem to describe all of observable reality. Yet, they are mutually contradictory. Therefore, they must be reflections of something far deeper, far more profound, clearly more fundamental to the nature of reality.

Expressed in another way, physicists continue to work diligently to find a way to unify general relativity (gravity) with quantum mechanics (including the standard model) to form a coherent whole. Both of these notions are supported by massive amounts of verifiable evidence. Both are clearly reflective of the fundamental nature of reality. Because they cannot be reconciled, however, we are left with the conclusion that they both **emerge** from something even more fundamental.

In a way, it's similar to the parable of the blind men trying to figure out what an elephant is by touch alone. Each blind man perceives some aspect of the elephant such as the trunk, the legs, the tail, the tusks, etc. but only with a unification of their reports can the elephant actually be defined.

One idea that attempts to find this unification or reconciliation is called "string theory." Stated another way, string theory attempts to develop a complete theory of gravity and quantum mechanics, a theory of quantum gravity, as it were. String theory is a purely mathematical, unified description of all forces and all masses, including gravity. It may be the theory of everything.

Mathematically, string theory postulates tiny strings, vibrating in different ways, as the fundamental basis of reality. As a result of their different vibrational modes, these strings define and create the various elementary particles and the forces of Nature. The vibrations of the tiny strings are highly reminiscent of the harmonics of the vibrating string of a cello that we described in Chapters 4 and 7. Just as the permitted vibrational modes (the standing waves) of an ordinary string are determined by its length and tension, these tiny strings vibrate in certain defined ways as well.

String theory, however, requires more dimensions than the four dimensions of space-time, with some of them "compactified" and therefore invisible for all practical purposes. The precise size and shape of the extra dimensions have a major effect on the physics and on the particular version of string theory postulated. The latest formulation of string theory, first described by Edward Witten in 1995, is called M- theory and leads to the mathematical notion of objects called branes.[9] M-theory also predicts the existence of multiple universes and is considered by many physicists to be the leading candidate for the theory of everything.

One of the successes of M-theory, as an expression of string theory, is that it embraces all past verifiable evidence of physics and continues to develop even further, making continuing contributions. The most important issue with M-theory is that it has not yet produced a prediction about Nature that can be tested. Hence, it's not yet science. Unequivocally, however, it is beautiful and useful mathematics. Do remember that mathematics has repeatedly led us to a better understanding of reality. Even if string theory does not turn out to be the theory of everything, it may well lead us forward. Just as music led Pythagoras to mathematics, and then to the idea that mathematics reflects the beauty of reality, any testable theory of everything will undoubtedly arise, at least initially, from mathematics.

Artwork provided by Glenn Miller, Clayton Photos

String theory is not the only idea that attempts to unify general relativity with quantum mechanics. One idea with a fairly long history is called **loop quantum gravity** and it is probably the most widely studied alternative to string theory.[10] Physicists have also developed several other ideas, one of which is that the Universe is actually a **hologram** or a **simulation**. Another is that the fundamental nature of reality is simply one thing: **information**. Yet another relies upon **geometry** to define the fundamental nature of reality, and this is referred to as "emergence theory," which postulates that reality is built from tetrahedrons (3-dimensional equilateral triangles) that in turn encode information in a quasicrystal, without any requirement for higher dimensions. While these and other ideas are being studied intensively by various groups of theoretical physicists, no one can know whether any will ever provide a theory of everything.

As physicists work diligently to evaluate all of these alternatives using the powerful sytem of thinking of scientific naturalism, some people have vehemently argued that these explanations are simply "voo-doo," no better than the supernaturalism of medieval Europe. The difference, however, is something that has demonstrated a profound track record of success: mathematics. No supernaturalistic idea has ever been expressed or predicted mathematically. In stark contrast, all of the notions described above about the fundamental nature of reality are essentially mathematical. While we may well be stumbling in the darkness of not knowing, mathematics reliably remains the headlamp that illuminates the path before us. We will undoubtedly make many wrong turns, as we do so often in science. But that's all part of the great human adventure of exploration.

A Recapitulation Before the Coda

Let's pause for a moment and summarize (recapitulate) before we go to the coda (in music, the ending or closing section) of our inquiry. We've discovered much together in this chapter:

1. We don't yet understand the fundamental nature of reality but we're getting closer, with the power of mathematics leading the way;
2. Einstein's theories of special and general relativity revolutionized our notions of time, space, mass, energy, and gravity and every bit of empirical evidence so far obtained supports those theories;
3. We've developed a standard model of elementary particles based both on mathematics and empiricism, although these particles are better thought of mathematically as ripples in fields;
4. Quantum mechanics, wherein the critical idea is that particles behave as either waves or particles depending on experimental conditions, is real, well-supported, but still not understood;
5. Physicists continue to struggle with the unification of general relativity's gravity and quantum mechanics to derive a theory of everything, that is, a theory of quantum gravity;
6. String theory (M-theory), the leading candidate for such a theory of everything, has its flaws and therefore several other approaches are being studied as well.

We've come very far since the beginning of the Scientific Revolution but there is so much else that we don't yet understand. We don't understand, for example, how light travels at 300 million meters per second, as opposed to some other finite speed. We don't understand how certain fundamental, ubiquitous constants, such as Planck's constant ($\hbar$) or pi (π), assume their particular values. We don't know how consciousness exists. We don't even understand for certain whether time is emergent or fundamental, even though physicists can measure time more accurately than anything else. It's all still mysterious. But such mystery wouldn't bother Einstein at all. On the contrary, he'd cherish it.

Waves: The *Ostinato* of Reality—The Persistent, Ubiquitous Motif of the Cosmos

You've probably noticed one particularly important thing by now. **Waves**, or more specifically, quantized vibrations, occur repeatedly and persistently—like a musical *ostinato*—in all of this physics—and throughout Nature itself. For example, elementary particles and atoms behave like waves until observed. Very recently (2016), scientists have developed the technology required to detect gravitational waves—ripples in the very fabric of space-time—and found them, just as the mathematics of Einstein's general relativity predicted. Elementary particles are best described mathematically as ripples in quantum fields. As we discussed above, string theory uses vibrational modes (standing waves) of tiny strings to define all elementary particles and all known forces, thereby mathematically aiming to unify gravity with quantum mechanics.

Brian Greene, a prominent string theorist and science communicator, says:

"Just as the strings of a cello can give rise to a rich variety of musical notes, the tiny strings in string theory vibrate in a multitude of different ways making up all the constituents of Nature. In other words, the Universe is like a grand cosmic symphony, resonating with all the various notes these tiny vibrating strands of energy can generate"

Another prominent string theorist, Michio Kaku, waxes even more poetically:

"So, the sub-atomic particles we see in Nature…are nothing but musical notes on a tiny vibrating string. What is physics? Physics is nothing but the laws of harmony that you can write on vibrating strings. What is chemistry? Chemistry is nothing but the melodies you can play on interacting vibrating strings. What is the Universe? The Universe is a symphony of vibrating strings."

And if this is actually the case, then we humans, as part of the Universe, not only **make** music, but we, ourselves, **are,** in our very essence:

Cosmic Music.

Notes

Chapter 1. What is Science?

1. Coyne (2015) provides an inclusive summary of various means of inquiry, along with their standards for the determination of plausibility. Science (scientific naturalism) stands alone in that it demands both falsifiability and predictability for an explanation to be judged plausible.
2. I choose these two theories with great care. One is from biology, the other from physics. Both have stood the test of time (a century or more). Both have produced many predictions that have consistently been demonstrated empirically. Both elegantly display the power of science and both have provided intellectual scaffolding for many other explanations.
3. This is a point even more important than it at first appears. Many people think that science offers "proofs." They will say: "Science **proves** such and such." It never does. It cannot do so because of the limitations of its inquiry. What it does do splendidly is to provide a universal standard by which an assertion can be judged to be plausible or implausible, often very highly plausible or very much implausible. This standard is universal, cross-cultural and global because it is based upon verifiable evidence. Scientists certainly argue about the quality of the evidence and of course they argue about its interpretation. They don't ever argue that consistent, replicated, verifiable evidence should be ignored, though sometimes such evidence is not appreciated until years after it is obtained. The concept of "truth" in science is useful for practical purposes when an explanation is extremely powerful, has been well tested many times, and is highly predictive. In that limited sense, both Darwin's and Einstein's theories are true.

Chapter 2. What is Music?

1. The goal here is to find a definition that is accurate, simple, direct, non-pretentious, non-metaphysical, and sufficiently general so as to be without cultural bias. Some might argue that the working definition here is too simple and doesn't reflect the nuances of their own particular notions about music, especially that it be "beautiful" or "inspiring" or "emotionally evocative." While I respect that point of view, I think that the definition meets our needs for this book very well, principally because it is so specific and concrete, neither narrow nor subjective, thereby enabling scientific examination. Cross-cultural studies show clearly that the determination of whether music is "beautiful," "inspiring," or "emotionally evocative" is in part culturally determined and is certainly a determination specific to an individual human being.
2. Our wrist watches contain a quartz crystal shaped like a tuning fork that when energized electrically vibrates at precisely 2^{15} times per second, that is, 32,768 times per second.

3. The theories of special and general relativity are discussed in much more detail in Chapter 15. Also refer to Greene (2003) and Kaku (1995).

Chapter 3. The Pulse

1. Circadian rhythms are one of the most universal characteristics of organisms. Such rhythms are found in bacteria, plants, as well as in animals. As the name implies, the rhythm occurs endogenously over an approximate 24-hour cycle, but will drift (usually forward) absent an external cue, which is commonly the light-dark cycle. The rhythm can be entrained externally and is corrected for changes in external temperature. The mechanism for maintaining the clock is highly conserved and involves three proteins, which have now been identified. In mammals, the clock resides in a small portion of the hypothalamus of the brain. Destruction of these cells results in complete disruption of the clock. The main point here is that endogenous time-keeping is biologically universal.
2. Jourdain (1997) makes a point of distinguishing between Gregorian chant and other forms of music with regard to the pulse. Even he, however, does not deny that the pulse exists but rather that it does not display "strict temporal patterning." (p. 125). As a musician, my experience is that the pulse is indeed the underlying organizational mechanism for every bit of music that I've ever performed, though its periodicity is not necessarily even or uniform. While Gregorian chant and other chant-like music performed by a solo artist may have a highly variable pulse, the pulse generally becomes far more uniform with the addition of other singers. The pulse is what keeps music coherent and aligned, though it is a challenge for any musician to maintain an absolutely constant internal pulse on his own. Interestingly, if the pulse is actually absolutely uniform, as with an electronic drum machine, it often sounds "artificial" and even unpleasant.

Chapter 4. Sound

1. Obviously, I've simplified the physics of sound a great deal, largely by leaving out the mathematics, and in part by omitting a great many background theoretical considerations, while preserving the fundamental concepts. In physics curricula, the study of sound is part of the study of simple harmonic motion. Such motion can also occur with an oscillating vertically suspended spring and weight or a swinging pendulum. Mathematically, these phenomena are described by either the sine or cosine, two trigonometric functions that reflect the periodicity of such motion. For those interested in the details of the mathematics, I suggest consulting Olson (1967), White & White (1980) or any good college physics text.
2. The period of a wave is defined as 1 / frequency, that is, the reciprocal of the frequency of the wave. The general equation for a sine wave is: $y(t)=A\sin(2\pi ft+\varphi)$ where A is the amplitude of the wave, π is, of course, the ratio of the circumference of a circle to its diameter, f is the frequency of the wave in cycles

per second, t is time, and φ is the phase of the wave, representing any shift from the point in the series of compressions and rarefactions where we begin the mathematical description.

3. A mathematical relationship useful for thinking about music is the relationship between the wavelength and the frequency of the sound wave. These are related to each other and to the speed of propagation of the wave by the relationship: $v=f\lambda$, where v is the speed of the sound wave, f is the frequency of the wave, and λ is the wavelength. From a practical, musical standpoint, the speed of sound in air depends on the temperature of the air and the mixture of gases in it, most relevantly, the content of water vapor (relative humidity), though this effect is much less than that of air temperature. The speed of sound increases significantly in warm air. For a wind instrument, λ is fixed, determined by the geometry of the instrument. So, if the speed of sound increases in warmer air, the frequency of the sound must increase, and the musician must adjust for that effect. Wind musicians know that they "go sharp" when the air temperature rises.
4. This simple algebraic equation is stated as follows: $f=(1/2L)\sqrt{T/\mu}$ where f is the fundamental frequency of the string, L is the length of the string, T is the tension on the string, and μ is the mass per unit length of the string. Translating to English, this equation means that all else being the same, a shorter string will produce a higher fundamental frequency (pitch), a tighter string will produce a higher fundamental, and a lighter string will produce a higher fundamental. Therefore, to get a low pitch, we'd use a long, loose, heavy string.
5. As further clarification of the perception of **beats** or **beatings** when two notes are "out of tune," I refer the reader to White & White (1980) p. 58. Beatings are perceived by the musician as a rapid fluctuation in the loudness of the sound. When two pitches are exactly "in tune," no beatings are heard because the crests (peaks) and troughs (compessions / rarefactions) of the wave forms coincide. As the two pitches become further "out of tune," that is they become more and more different in frequency, a periodic alignment and counter-alignment of the peaks (crests) and troughs of the waves occurs, resulting in the sensation of beatings. More precisely, this is described as recurring constructive and destructive interference. The frequency of the beatings is determined by the equation: $f_{\text{Beating Frequency}}=f_2-f_1$ where f_2 is the higher frequency and f_1 is the lower frequency. The greater the difference between f_2 and f_1, the more frequent the beatings become. As the difference increases from zero, initially the beatings are so slow as to be barely perceptible, but they gradually increase in frequency. Musicians divide the difference in frequency between two semitones (see Chapter 5) into 100 "cents." Generally, professional musicians can barely tolerate a disparity of 3 cents when playing together in unison (playing the same pitch); more than that becomes increasingly unpleasant. For professional musicians, a difference of 10 cents is completely unacceptable. As the frequency difference of the two pitches approaches the distance of half a semitone (50 cents either sharp (up) or flat (down)), they become perceived as two dissonant notes rather than as a single "out of tune" note. A further complication is that even if the fundamental frequency is fairly well in tune, some of the harmonics may not be, and this can also

result in the perception of beatings. Yet another complication arises when musicians play in harmony and must listen to both the fundamental frequencies and the interactions of their harmonics. Intonation, therefore, must always be somewhat of a compromise between the "perfect" and the "musically good enough under the circumstances."

6. Often harmonics (more accurately, harmonic partials) and overtones are confused. They are different. A harmonic is defined as a whole number (integer) multiple of the fundamental pitch, as described in the text. In contrast, an overtone is any frequency greater than the fundamental, regardless of whether or not it is an integer multiple. Such an overtone would be referred to either as a harmonic partial or an inharmonic partial. The nomenclature can be even more confusing. When speaking of harmonic partials, the fundamental is the first harmonic, the second harmonic is the fundamental x 2, etc. But, when speaking of overtones, the fundamental is referred to as simply the fundamental, the first overtone is the fundamental x 2, the second overtone is the fundamental x 3, etc. Finally, percussion instruments often produce inharmonic partials. In this book, for the sake of minimizing confusion, I refer only to harmonics, realizing that the nomenclature may thereby be slightly inexact.
7. The relative strength of the various perceived harmonics of a vibrating string are much influenced by the shape and type of resonator (see Chapter 7). The body of a violin, for example, reinforces some harmonics and softens other harmonics of the vibrating string, resulting in a characteristic mixture of intensities of those harmonics. Therefore, even if all harmonics are present in the vibrating string, not all may be heard in the sound emanating from the instrument. In addition, the violin body is of course vibrating sympathetically with the violin string, thereby contributing its own mix of harmonic or inharmonic partials. This phenomenon similarly results from the shape and size of the vocal apparatus of humans, which allows us to distinguish one voice from another. For those particularly interested in the topic of harmonics, I recommend a simple experiment. If you have access to an acoustic piano that has been tuned recently, gently depress the middle C key to lift the damper from the piano string(s), and then sound the C below middle C loudly and briefly. Listen. You will hear the middle C string(s) vibrating in sympathy with the second harmonic of the note that you just sounded. If you have a good ear, you can do the same thing with the G above middle C, and hear the third harmonic of the original piano string(s).

Chapter 5. Melody

1. The **tune** is a term most often used to refer to an entire melody, with its beginning, middle, and end. It is also a term used most frequently to describe a sequence of pitches that is memorable and can be sung easily. Melodies don't always have those characteristics, especially in Western classical music. Indeed, they may be quite fragmentary, and be referred to as **themes** which may or may not be sung easily at all. Finally, just a few notes in succession is often called a **motif**. Motifs can be parts of themes,

or even parts of melodies in the sense of a complete tune, or they can stand entirely on their own as an important musical statement.

2. Simplistically, the octave could be divided into very small intervals, differing by only 1 Hz, which seems to be the limit of human perception of difference in pitch. Practically, however, as a result of both physics and brain biology, this doesn't happen, with the possible exception of one instrument, the electronic Theremin, named after its inventor. The ordinary, practical limit appears to be about 24 distinct intervals. To Western ears, accustomed as we are to 12 equal intervals dividing the octave, such a scale seems to be "microtonal" and exotic. From the standpoint of music around the globe, 12 seems to be a popular division.
3. Musical notation has evolved markedly during the last 10 centuries, increasing in complexity and accuracy. While printed notation of course tends to "fix" musical interpretation, especially when well annotated, contemporary high-fidelity recordings of performances have "locked in" certain widely enjoyed interpretations of the standard repertoire as well. This has both advantages and disadvantages.
4. In writing these sections on music theory, I wrestled about whether to print written Western music out as it's normally notated, with stave lines, etc. Because this is a book about music and science, I very much wanted it to be completely accessible even by those who had never learned to read written notation at all, which by the way includes some very talented and famous American pop singers. With the availability of online pianos, it became obvious to me that illustrating a keyboard and providing notation simply as letters of the alphabet would do nicely, avoiding the necessity for learning and reading Western notation, even though the rhythm of the notes could not be communicated. By using online resources, every melody or chord progression in this book can easily be accessed and heard.
5. Wind instrumentalists can adjust their intonation somewhat to achieve better harmonies. They do this by tightening or loosening their embouchures (the shape of their lips and oral cavities as they interact with the mouthpiece of the instrument) to raise or lower the pitch slightly. They also often have a choice of "fingerings" to produce particular pitches. Professional musicians make constant, sometimes very subtle alterations to improve the sound of the ensemble. To hear the extent of possible pitch alteration in a fixed-pitch instrument, simply listen to the opening clarinet *glissando* (technically really a smear) from George Gershwin's *Rhapsody in Blue.*
6. Several systems of just intonation exist, each generated by choosing particular ratios of the pitches of the scale to each other. Musicians must constantly make compromises by deviating slightly from the simplest numerical ratios, which tend to be the most harmonious. Equal temperament is the most commonly used system for large ensembles, though string players and vocalists will often use a combination of standard Pythagorean and just intonation when performing with each other, depending on what "works" best.

7. Herman von Helmholtz was a 19th century physicist interested in many aspects of science, one of which was sound. In particular, he was interested in the physical principles of musical sound production and the underlying physical basis for harmony. See Helmholtz (1954).
8. I fully realize that many other scales have been used in Western music, including a range of diatonic scales (modes) given euphonious Greek names, as well as whole tone scales, random chromatic tone rows, etc. And I realize that there is a fourth type of minor scale as well. The ordinary major scale is also called the Ionian mode (beginning on C and playing only the white keys of the piano) and the natural minor scale is also called the Aeolian mode (beginning on A and playing only the white keys of the piano). The less often used modes, the Dorian, Phrygian, Lydian, Mixolydian, and Locrian can all be played using the white keys of the piano by starting at other pitches of the C-major scale. The Dorian begins on D, the Phrygian on E, the Lydian on F, the Mixolydian on G, and the Locrian on B. The consideration of only major and minor scales herein is therefore a simplification, but a useful one, especially as we discuss the psychology of music in later chapters. No law says that the only good melodies are those derived from the diatonic scale, nor that they must be tonal, nor that they must end on the tonic. Indeed, the opposite is the case. Simply listen to any Debussy composition based on a whole tone scale.

Chapter 6. Harmony

1. The history of music is a fascinating, complex study. I regret that it receives so little attention in this book, and then only enough to provide some kind of frame for the discussion of music theory and human musicality. For those who do not already possess this historical framework, I recommend Hanning (2002) as an excellent introduction.
2. The addition of the major or minor third interval to form a triad was far more revolutionary than it appears to us contemporary humans. Like most changes in accepted harmonic conventions, it probably "edged" its way in gradually. That's certainly the case for complex jazz chords which would have sounded horribly dissonant to our ancestors, even to very sophisticated composers. The history of Western music is a history of greater and greater harmonic complexity within a tonal framework. Interestingly, atonal, polytonal, minimalist, or highly chromatic music still has not achieved widespread popularity among even the fairly sophisticated listening public, except perhaps for brief exposures in the concert hall. This, too, could change. Additionally, homophony, while familiar now to most of the cultures of the world, rarely developed independently in non-Western cultures. Melody with an underlying chord structure, as opposed to emphasis on melody / rhythm, appears not to develop as a routine part of musical evolution universally.
3. Helmholtz (1954), as translated from the last German edition, published in 1877. Helmholtz made numerous contributions to physics in a variety of areas. His work in acoustics and musical aesthetics remains influential as a basis for thinking about some aspects of music theory.

4. Fishman, Yonatan I., Igor O. Volkov, M. Daniel Noh, P. Charles Garell, Hans Bakken, Joseph C. Arezzo, Mattew A. Howard, and Mitchell Steinschneider. 2001. "Consonance and Dissonance of Musical Chords: Neural Correlates in Auditory Cortex of Monkeys and Humans" *J. Neurophysiology* 86: 2761-2788.
5. Bowling, Daniel L. and Dale Purves. 2015. "A Biological Rationale for Musical Consonance." *Proceedings of the National Academy of Science USA* 112(36): 11155-11160.
6. Cousineau, Marion, Josh H. McDermott, and Isabelle Peretz. 2012. "The Basis of Musical Consonance As Revealed by Congenital Amusia." *Proceedings of the National Academy of Science USA* 109(48): 19858-19863.
7. One vocal group that I can recommend heartily is called **The King's Singers**. Many of their performances are available online. They use arrangements with astonishingly beautiful chord progressions and sing with near perfect intonation.

Chapter 7. Making Music

1. Much of this material is very well described and analyzed by Benade (1990 and1992). For those interested in a detailed mathematical discussion of the physics and acoustics of musical instruments, I certainly recommend both of these books. Powell (2010) presents a much lighter, non-mathematical description of some of the same material.
2. The description of the instruments of the orchestra has been approached in numerous ways by various authors. For this discussion about musicians and their instruments, I rely a great deal on my personal experiences in performing with many different types of bands, orchestras, and small ensembles. I also thank Tina Ward, clarinetist with the St. Louis Symphony, for her critical reading of this chapter.
3. Because of its simplicity, the canjo is an excellent tool to demonstrate the principles of musical acoustics. It can be purchased online at reasonable cost.
4. Adrian Cho. 2010. "Probing the Secrets of the Finest Fiddles." *Science* 328: 1468-1470.
5. Making a soda (or drinking) straw oboe is a pleasurable and instructive musical activity to do with children ages 4 to 8 or so. Even many adults unfamiliar with musical acoustics find this to be fun. In class, I was surprised at the delight that a group of older adults took in "making" a musical instrument on their own. The drinking straw oboe is also described in Powell (2010) p. 73, and instructions and diagrams are readily available online.
6. The concept of the standing wave is important in acoustics. These are also referred to as stationary waves and can occur in strings or in enclosed columns of air. We've already seen diagrams of standing waves on strings (Chapter 4) when we discussed sound. The idea is quite simple conceptually. When the material (the string or air in a column) is set in oscillatory motion, a wave is created with apparently stationary peaks (crests) and troughs. Within that stationary wave, we observe nodes and

antinodes. In a pipe filled with air, the nodes are points in the wave where the molecules appear not to move, whereas the antinodes are the peaks and troughs of the standing wave. The placement of nodes and antinodes is a function of whether the column is open or closed (at one end) to the atmosphere as well as its shape (conical or cylindrical). The length of the standing wave is determined by the length of the column or the string. Because only a node can occur at the closed end of a cylindrical pipe (like a clarinet), such a tube can produce only the odd harmonics (1x, 3x, 5x, etc.) whereas in a cylindrical pipe open to the atmosphere at both ends (like a flute), all harmonics are represented because there exist antinodes at both ends. In contrast, a closed conical pipe, *e.g.* a saxophone, behaves very much like an open cylindrical pipe, like a flute. For those interested in a complete physical and mathematical description, see White & White (1980).

7. Montague, Michael and Tina Ward. 2012. "Reeds: Good or Bad? It's in the Cane—An Inside View of *Arundo donax* L." *The Clarinet* 40, no. 1 (December): 62-65.
8. Another difference between a clarinet and a flute (and a soprano saxophone) is the position in the musical scale where the instrument "overblows." A flautist can play a low D (the D just above middle C) on her instrument and then with a slight change of embouchure (the position of her lips and jaw), "overblow" to the octave above, maintaining exactly the same finger position. Note that the octave is the second harmonic, and it is even (2x). With the clarinet, because it is a closed cylindrical pipe, producing predominantly the odd harmonics (1x, 3x, 5x, 7x, etc.). the musician overblows instead to the third harmonic (3x) which is a 12th above the fundamental. Thus, the physical acoustics of the two types of instruments have a major influence on the capabilities of the instruments.

Chapter 8. The Ear

1. Shubin (2009) provides a nice summary of this history, starting in the 19th century with embryology even before Darwin, and describes a growing realization among scientists that two bones of the mammalian middle ear were once bones in the jaws of reptiles.
2. To minimize the number of terms that I use in describing genetics and evolution, I will routinely refer to "genes" where a better term would sometimes be "alleles." Genes can exist in various forms, which we refer to as alleles. The trait conferred by the presence of one allele of a gene may be different from that conferred by another allele of the same gene.
3. If I were asked to provide a *coup de grâce* to an assertion that biological simplicity cannot possibly give rise to complexity without a "designer," I would describe the molecular biology that demonstrates the very close homology (similarity) of genes of all life forms on earth and that the closer such forms are related evolutionarily, as judged paleontologically, anatomically, physiologically, and taxonomically, the more homologous are their genes. In addition, I'd explain the concept of gene duplication, followed by divergence, which provides a very simple mechanism for increasing complexity. As a reply to an argument about the necessity of "design" for "improvement" in organisms, I'd simply provide

verifiable evidence that our human bodies are very poorly "designed" with numerous stupidities of anatomy. One example: what engineer would design a gland to surround the male urethra, when said gland was destined to enlarge as the man aged? The name of that gland? The prostate. Nature tinkers well; it designs not at all. In short, no "designer" or "guiding force" of any kind is required for the evolution of organisms by natural selection or for the emergence of complex forms from simpler forms. It is the process of natural selection itself that forces the direction of change. This concept was Darwin's insight of genius.

4. One of the criticisms aimed at discrediting the theory of evolution by natural selection is the demonstrably false assertion that "there are no transitionary forms to be found in the fossil record." This statement is simply and completely in error. Along with an untold number of similar examples, abundant transitionary life forms have been identified by paleontologists to support the re-purposing of the jawbones of early mammal-like reptiles to become bones of the mammalian middle ear.
5. Urban, Daniel J., Neal Anthwal, Zhe-Xi Luo, Jennifer A. Maier, Alexa Sadler, Abigail S. Tucker, Karen E. Sears. (2017) "A new developmental mechanism for the separation of the mammalian middle ear ossicles from the jaw. *Proceedings of the Royal Society B: Biological Sciences.* 284 :(1848).
6. Carter (2009) presents beautiful illustrations and a functional description of the human ear.
7. "Orders of magnitude" constitute a short-hand method that scientists use when numbers vary over a wide range. They are commonly expressed as exponents. For example, $10^1 = 10$; $10^2 = 100$; $10^3 = 1000$, $10^4 = 10{,}000$, etc. Similarly, $10^0 = 1$; $10^{-1} = 0.1$; $10^{-2} = 0.01$; $10^{-3} = 0.001$, etc. Hence, the exponent reflects the order of magnitude. Amazingly, the human ear is capable of perceiving sound over 12 to 13 orders of magnitude, depending on the frequency, in large measure due to the repurposing of those reptilian jaw bones.

Chapter 9. Our Musical Brain

1. Sheet lightning is one type of lightning, all types of which occur because of the formation of a channel of ionized air molecules that carry an electrical discharge from one area of electric charge to another. Sheet lightning occurs within a single cloud, with the spark remaining within that cloud. It's actually the most common type of lightning and gets its name from the fact that it illuminates the sky with a "sheet" of light. Here, I'm using it as a metaphor for the patterns of electrical activity within the brain.
2. "There isn't a cognitive function that doesn't somehow pertain to music. People have realized that music really does serve as a gateway into understanding human cognition." Robert Zastorre, neuroscientist. As quoted in Mannes (2011) p. 33.
3. One example of this change has been reported by several independent investigators: an increased size of the corpus callosum, which is a "cable" of myelinated axons that connect the left and right hemispheres of the brain. It is larger in the brains of musicians than in those of non-musicians. The result

is credible because of replications with different populations. See: Schalaug, G., L. Jancke, Y. Huang, J.F. Staiger, and H. Steinmetz. 1995. "Increased corpus callosum size in musicians" *Neuropsychologia* 33(8): 1047-55. For general observations about the difference in brain structure between musicians and non-musicians, see: Gaser, c., and G. Schlaug. (2003). "Brain structures differ between musicians and non-musicians. *The Journal of Neuroscience* 23(27): 9240-9245.

4. Although very nice illustrations of the brain are available online, I highly recommend Carter (2009), not only for its illustrations, but for its concise text. While not inexpensive because of the quality of the illustrations, it's a very impressive "coffee table book" that will quite astonish your friends, and teach them a lot too. Most of the discussion herein about both the microscopic and macroscopic anatomy of the brain is derived conceptually from this reference. I've deliberately simplified the discussion and descriptions, however. A common, much more rigorous way to think about the overall anatomy of the brain is to divide the organ into anatomical parts, *viz.* the Prosencephalon (Telencephalon + Diencephalon), the Mesencephalon, and the Rhombencephalon, described as follows:
    a. The **Prosencephalon** or Forebrain, consisting of the cerebrum, with two-thirds of the brain's mass;
    b. The Prosencephalon consists in turn of two subdivisions.
        i. First is the **Telencephalon** (the cerebral cortex with its four lobes, frontal, parietal, occipital, and temporal), each lobe represented in both hemispheres of the cerebrum. The Temporal lobe houses part of the limbic system, including the amygdala and the hippocampus.
        ii. Second, the Prosencephalon also encompases the **Diencephalon**, which in turn consists of limbic system structures including the thalamus, the hypothalamus, and the pineal gland
    c. The **Mesencephalon** or Midbrain connects the Forebrain to the **Rhombencephalon** (Hindbrain). The Midbrain regulates movement and aids in the processing of visual and auditory information with structures that include the tectum, the cerebral peduncle, and the substantia nigra.
    d. The Hindbrain regulates autonomic functions and includes the metencephalon (the pons and cerebellum) and the myelencephalon (the medulla oblongata).
    e. The Brainstem is usually considered to be composed of the combination of Mesencephalon and Rhombencephalon.

5. The aim here is to provide the basics of neuroscience, neurophysiology, and neurochemistry accessibly, of course with a particular focus on music. Much of the basic information is also available in Guyton (1992).

6. Neurotransmitters are small molecules like serotonin, norepinephrine, acetylcholine, and dopamine. Drugs used for the treatment of depression, psychosis, and other brain diseases, as well as to enhance cognition, often modify the levels of and / or the sensitivity of neurons to these neurotransmitters.
7. Bertolero, Maxwell A., B. T. Thomas Yeo, and Mark E'Esposito. 2015. "The modular and integrative functional architecture of the human brain." *Proceedings of the National Academy of Science U.S.A.* 112(49): 6798 – 6807.
8. Sachs (2007) p.313.
9. Blood, A. J. and R. J. Zatorre. 2001. "Intensely Pleasurable Responses to Music Correlate with Activity in Brain Regions Implicated with Reward and Emotion." *Proceedings of the National Academy of Sciences U.S.A.* 98:11818 – 23.
10. Thaut, Michael H., Pietro Davide Trimarchi, and Lawrence M. Parsons. 2014. "Human Brain Basis of Musical Rhythm Perception: Common and Distinct Neural Substrates for Meter, Tempo, and Pattern." *Brain Science* 4(2): 428-452.
11. Harris, Robert, Bauke M. de Jong. 2015. "Differential parietal and temporal contributions to music perception in improvising and score-dependent musicians, an fMRI study." *Brain Research.* 1624: 253-264.

Chapter 10. Music Changes Us

1. The main sources for this section are academic texts: Huron (2007) and Radocy & Boyle (2012). I've endeavored to extract a more accessible narrative of some of the fundamental ideas from these texts and other sources, especially Ockelford (2017). Also helpful at a more aesthetic level is Storr (1992).
2. The experiment described in detail in this section comes from the work of Stefan Koelsch (2013) p. 110. Most of the information in this section of Chapter 10 is derived from that reference. For those interested in much more scientific detail, I highly recommend his book, as well as the original publications of Koelsch and other neuroscientists in its reference list. That list alone is well worth the price of the book because the list displays the immense amount of attention that neuroscientists are giving to the musical brain. I've chosen this particular experiment for exposition because it is representative of a whole class of similar protocols. Another clever experimental design has been used to examine the brains of highly skilled musicians in the process of improvisation by providing a small keyboard, together with appropriate mirrors and earphones, within an fMRI machine.

Chapter 11. Music Heals Us

1. Stegemolier, Elizabeth. 2018. "Parkinson's Disease Patients' Singing Improves Voice and Swallow Impairment. *Today's Geriatric Medicine* 11(1): 18.

2. The bulk of basic information in this chapter is derived from more academic texts. Chief among these is Bruscia (2006), which is a standard text for all music therapists. With a different, more intuitive and less data-driven approach, Stevens (2012) is also helpful conceptually and practically. I've endeavored to create a concise, pithy narrative of this information about music therapy in a single chapter, rather than in an entire book, as those authors have accomplished so well. Still, the key points are all here, as are the scientific underpinnings.
3. While the references listed in note (2) above are very useful and extremely well documented, a much more accessible source is: Thompson, William Ford and Gottfried Schlaug. (2015) "The Healing Power of Music: New Therapies Are Using Rhythm, Beat and Melody to Help Patients with Brain Disorders Recover Language, Hearing, Motion, and Emotion." *Scientific American Mind*. March / April.
4. Koelsch, Stefan. 2009. "A Neuroscientific Perspective on Music Therapy." *Annals of the New Yourk Academy of Sciences*. 1169: 374-384.
5. Gaynor (1999). This is a well-written book that emphasizes the underlying bases for the effects of music in clinical medicine, as opposed to how a music therapist should actually practice her or his craft. It's the best and most accessible summary that I've found for information about the physiological effects of music on the human body, though much more has been learned since its publication.

Chapter 12. The Composer's Toolkit

1. Borodin is a rare example of a composer / scientist polymath. While it's not at all uncommon for scientists and mathematicians to be good musicians and even to perform professionally, the composition of memorable and enduring music seems to require a special skill set, a type of genius, and / or a level of attention that a serious scientist almost never can give, especially if he or she is to maintain scientific credentials and to compete in the world of scientific research. Borodin's biography is available in many places online or in references such as Cross & Ewen (1962) and Abraham (1927).
2. YouTube
3. The tools and techniques of composition have been analyzed extensively by musicologists and theorists. The basic concepts are found in several books of the bibliography here with self-evident titles. Ultimately, however, the successful composition of enduring music seems to require a certain as yet inexplicable talent. This is part of the remaining mystery of music. It's certainly possible to analyze a composition and to enumerate the "tricks" and devices that a composer has employed. But the explanation for the difference between a Mozart and a Salieri, even though both wrote in much the same style with many of the same conventions, seems ineffable. Mozart's musical ideas were just so much better and what he could do with those ideas, seemingly effortlessly, was nothing short of astonishing. Finally, remember that it's not uncommon for the work of a composer to be much more appreciated decades after the composer's death.

4. Huron (2007) has written an entire volume with documentation to support these ideas.
5. Among the best of these is the widely recognized textbook by Adler (2002).
6. See Levitin (2006) for a nice discussion of the "pleasures" of frustrating or delaying the resolution of tension and how often it's been used by composers.

Chapter 13. Our Musical Humanity

1. Probably no one has done more to consolidate and to articulate the notion of complex vocal learning than Aniruddh Patel of Tufts University. He has also been an articulate proponent of the evolutionary relationship between music and language. Two key references: Patel (2008) and Patel, (2015).
2. Webb, D. M. and J. Zhang. 2004. "FoxP2 in Song-Learning Birds and Vocal-Learning Mammals." *Journal of Heredity* 96(3): 212-216.
3. Snowball is a YouTube superstar.
4. Mingle, Morgan E., Timothy M. Eppley, Matthew W. Campbell, Katie Hall, Victoria Horner, and Fran B. M. de Waal. 2014 "Chimpanzees Prefer African and Indian Music Over Silence". *Journal of Experimental Psychology: Animal Learning and Cognition* 40(4): 502-505.
5. Cihodariu, Miriam. 2011. "A Rough Guide to Musical Anthropology." *Journal of Comparative Research in Anthropology and Sociology.* 2(1): 2068.
6. Marcus, Gary F. 2012. "Musicality: Instinct or Acquired Skill?" *Topics in Cognitive Science.* 2012: 1-15.
7. Zatorre, Robert. 2016. "Amazon Music." *Science* 535: 496-497.
8. Conard, Nicholas J., Maria Malina & Susanne C. Munzel. 2009. "New Flutes Document the Earliest Musical Tradition in Southwestern Germany. *Nature.* 460: 737-740.
9. Also see notes 2 and 3 of Chapter 8 "The Ear."
10. Sachs (2007). pp: 98-119. Amusia or musical aphasia can occur in many different forms with a variety of deficits and apparent origins. The point of this discussion is that some forms apparently have a genetic component, suggestive that musicality *per se* has an inherited component also.
11. Darwin, Charles (1991). *The Descent of Man and Selection in Relation to Sex.* London: John Murray.

Chapter 14. Our Musical Children

1. Sachs (2007) Chapter 9.
2. Roden, Ingo, Dietmar Grube, Stephan Bongard, and Gunter Kreutz. 2014. "Does Music Training Enhance Working Memory Performance? Findings from a Quasi-experimental Longitudinal Study." *Psychology of Music.*42(2): 284-298.

3. Ho, Yim-Chi, Mei-Chun Cheung, and Agnes S. Chan. 2003. "Music Training Improves Verbal but Not Visual Memory: Cross-Sectional and Longitudinal Explorations in Children." *Neuropsychology* 17(3): 439-450.
4. Tierney, Adam T., Jennifer Krizman, and Nina Kraus. 2015. "Music Training Alters the Course of Adolescent Auditory Development." *Proceedings of the National Academy of Sciences.* 112(32): 10062-10067.
5. Habibi, Assal, Antonio Damasio, Beatriz Ilari, Ryan Veiga, Anand A. Joshi, Richard M. Leahy, Justin Pl. Haldar, Divya Varadarajan, Chitresh Bhushan, and Hanna Damasio. 2017. "Childhood Music Training Induces Change in Micro and Macroscopic Brain Structure: Results from a Longitudinal Study. *Cerebral Cortex* doi:10.1093/cercor/bhx286.

Chapter 15. Cosmic Music

1. This is indeed a very brief narrative covering 500 years. It's like fast-forwarding through history and seeing only the most pertinent highlights. For a far more detailed, cleverly presented history of the interaction between science and music, see Pesic (2014).
2. James (1993) and Pesic (2014) present highly coherent and insightful expositions of Greek mathematics inspired by music.
3. On page 37, James (1993) writes: "Pythagoras's discovery of the arithmetical basis of the musical intervals was not just the beginning of music theory: it was also the beginning of science."
4. Naturalism is a branch of philosophy that concludes from observation that there is nothing supernatural and that the natural is all that exists. Scientific naturalism is sometimes referred to as methodological naturalism, though I think that this term was created by those who don't really understand that the purpose of science is not to engage with a method but rather to explain, which is the goal of all philosophy. Some philosophers, for whatever reason, have created a pejorative term called "scientism" but I've not seen any of them clearly distinguish "scientism" from scientific naturalism, the underlying philosophy of modern science. Scientific naturalism does not automatically assume or presume (as a given) that nothing at all supernatural exists or can exist. If verifiable evidence for supernatural phenomena were to arise, scientific naturalism would immediately examine it and if the evidence met the standards of science, theories would change very quickly indeed. At least so far, scientific naturalism has found that the Universe (Multiverse, Cosmos) exists and behaves precisely, completely, predictably, and reliably as we would expect absent anything at all supernatural. But, as always in science, this is a provisional conclusion, subject to new or additional verifiable evidence, should it arise.
5. See also notes 2 and 3 from Chapter 8.
6. The remainder of this chapter relies heavily upon Greene (2003) and Kaku (1995).

7. More precisely, mass is a property that energy exhibits, but the details of explaining such a concept are beyond the purview of this book.
8. The standard model is certainly useful, but it's also incomplete when described only in this way. Much closer to the fundamental nature of reality is that these particles are actually best described as fields. Think of fields as a bit like constantly changing and rippling "fluids" everywhere in the Cosmos. The Higgs boson is then a field which confers mass. What we think of as particles are really waves of these fields.
9. The term "brane" comes from the notion that our entire space-time exists on a membrane "floating" through multiple universes. Please note that I've deliberately not discussed many other mysteries in theoretical physics and cosmology, such as black holes, dark energy, dark matter, the so-called big bang, anti-matter, the quantum foam, the expanding Universe, etc. I've done my best here to focus on those aspects of our exploration of the fundamental nature of reality that we experience in our earth environment, and especially those fairly directly related to music. For those interested in cosmology, the origin of the Universe, and similar very interesting topics, you will find many books by brilliant physicists, including Stephen Hawking. Black holes in particular are important because they constitute a singularity where quantum mechanics and gravity come together, just as at the big bang. All of these topics are interrelated but it's impossible to discuss them all in a single, accessible chapter.
10. See Rovelli (2016).

Books

History of Science

Pertinent to Chapters 1 and 15

Coyne, Jerry A. *Faith vs. Fact: Why Science and Religion are Incompatible.* New York, New York. Penguin Books. 2015.

Greene, Brian. *The Elegant Universe: Superstrings, Hidden Dimensions, and the Quest for the Ultimate Theory.* New York, New York. W. W. Norton & Company. 2003.

James, Jamie. *The Music of the Spheres: Music, Science and the Natural Order of the Universe.* New York, New York. Copernicus, Springer-Verlag, 1993.

Jeans, James. *Science & Music.* New York, New York. Dover Publications, Inc. 1968.

Kaku, Michio and Thompson, Jennifer. *Beyond Einstein: The Cosmic Quest for the Theory of the Universe.* New York, New York. Anchor Books. 1995.

Mannes, Elena. *The Power of Music: Pioneering Discoveries in the New Science of Song.* New York, New York. Bloomsbury. 2011.

Pesic, Peter. *Music and the Making of Modern Science.* Cambridge, Massachusetts. The MIT Press. 2014.

Rovelli, Carlo. *Seven Brief Lessons on Physics.* New York, New York. Riverhead Books. 2016.

History of Music

Pertinent to Chapters 2 and 12

Abraham, Gerald E. H. *Borodin, The Composer & His Music.* London. William Reeves. 1927.

Crocker, Richard L. *A History of Musical Style.* New York, New York. Dover Publications, Inc. 1986.

Cross, Milton and David Ewen. *The Milton Cross New Encyclopedia of the Great Composers and Their Music.* Volume 1. Garden City, New York. Doubleday & Company, Inc. 1969.

Hanning, Barbara Russano. *Concise History of Western Music.* New York, New York. W. W. Norton & Company. 2002.

Hindley, Geoffrey (ed.). *The Larousse Encyclopedia of Music.* New York, New York. Excalibur Books. 1984.

Machlis, Joseph. *The Enjoyment of Music.* New York, New York. W. W. Norton & Company, Inc. 1970.

Music Theory, Composition, and Performance

Pertinent to Chapters 3, 5, 6, 7, and 12

Adler, Samuel. *The Study of Orchestration.* 3rd Edition. New York, New York. W. W. Norton & Company, Inc. 2002.

Barlow, Harold and Morgenstern, Sam. *A Dictionary of Musical Themes.* New York, New York. Crown Publishers, Inc. 1975.

Benade, Arthur H. *Horns, Strings, & Harmony.* New York, New York. Dover Publications, Inc. 1992.

Black, Dave and Gerou, Tom. *Essential Dictionary of Orchestration.* Van Nuys, California. Alfred Publishing Co., Inc. 1998.

Gorow, Ron. *Hearing and Writing Music: Professional Training for Today's Musician.* Studio City, California. September Publishing. 2009.

Harder, Paul O. *Harmonic Materials in Tonal Music: A Programmed Course. Parts I and II.* Boston, Massachusetts. Allyn and Bacon, Inc. 1980.

Ricigliano, Daniel A. *Melody and Harmony in Contemporary Songwriting.* New York, New York. Donato Publishing Co. 1978.

Surmani, Andrew, Surmani, Karen Farnum, and Manus, Morty. *Alfred's Essentials of Music Theory.* Van Nuys, California. Alfred Publishing Co., Inc. 2004.

Wyatt, Keith and Schroeder, Carl. *Harmony and Theory: A Comprehensive Source for All Musicians.* Milwaukee, Wisconsin. Hal Leonard Corporation. 1998.

Physics of Sound and Music

Pertinent to Chapters 4 and 7

Benade, Arthur H. *Fundamentals of Musical Acoustics.* New York, New York. Dover Publications. 1990.

Helmholtz, Hermann. *On the Sensations of Tone as a Physiological Basis for the Theory of Music.* New York, New York. Dover Publications. 1954.

Olson, Harry F. *Music, Physics, and Engineering.* New York, New York. Dover Publications. 1967.

White, Harvey E. and White, Donald H. *Physics and Music: The Science of Musical Sound.* Mineola, New York. Dover Publications. 1980.

Biology of the Perception and Processing of Music

Pertinent to Chapters 8, 9, 10, 11, 12, and 14

Ball, Philip. *The Music Instinct: How Music Works and Why We Can't Do Without It.* Oxford, England. Oxford University Press. 2010.

Bruscia, Kenneth E. (ed.). *Case Studies in Music Therapy.* Gilsum, New Hampshire. Barcelona Publishers. 2006.

Carter, Rita. *The Human Brain Book: An Illustrated Guide to its Structure, Function, and Disorders.* London. Dorling Kindersley Ltd. 2009.

Gaynor, Mitchell L. *The Healing Power of Sound: Recovery from Life-Threatening Illness Using Sound, Voice, and Music.* Boston, Massachusetts. 2002.

Guyton, Arthur C. *Basic Neuroscience: Anatomy & Physiology.* 2nd Edition. Philadelphia, Pennsylvania. W. B. Saunders & Company. 1992.

Huron, David. *Sweet Anticipation: Music and the Psychology of Expectation.* Cambridge, Massachusetts. The MIT Press. 2007.

Jourdain, Robert. *Music, the Brain, and Ecstasy: How Music Captures Our Imagination.* New York, New York. HarperCollins Publishers. 1993.

Koelsch, Stefan. *Brain & Music.* Hoboken, New Jersey. John Wiley & Sons, Ltd. 2013.

Levitin, Daniel J. *This Is Your Brain on Music: The Science of a Human Obsession.* New York, New York. Penguin Group. 2006.

Ockelford, Adam. *Comparing Notes: How We Make Sense of Music.* London. Profile Books. 2017.

Patel, Aniruddh D. *Music and the Brain: Course Guidebook.* Chantilly, Virginia. The Great Courses. 2015.

Powell, John. *How Music Works: The Science and Psychology of Beautiful Sounds from Beethoven to the Beatles and Beyond.* New York, New York. Little, Brown and Company. 2010.

Radocy, Rudolf E. and Boyle, J. David. *Psychological Foundations of Musical Behavior.* Springfield, Illinois. Charles C. Thomas Publisher, LTD. 2012.

Sacks, Oliver. *Musicophilia.* New York, New York. Alfred A. Knopf. 2007.

Shubin, Neil. *Your Inner Fish: A Journey into the 3.5-billion Year History of the Human Body.* New York, New York. Vintage Books. 2009.

Stevens, Christine. *Music Medicine: The Science and Spirit of Healing Yourself with Sound.* Boulder, Colorado. Sounds True. 2012.

Storr, Anthony. *Music and the Mind.* New York, New York. Ballantine Books. 1992.

Thaut, Michael H. *Rhythm, Music, and the Brain: Scientific Foundations and Clinical Applications.* New York, New York. Rootledge, Taylor, & Francis Group. 2005.

Anthropology, Ethology, Evolution, and Music

Pertinent to Chapter 13

Lane, Nick. *Life Ascending: The Ten Great Inventions of Evolution.* New York, New York. W. W. Norton & Company, Inc. 2009.

Patel, Aniruddh D. *Music, Language, and the Brain.* New York, New York. Oxford University Press. 2008.

Music Education

Pertinent to Chapter 14

James, Abel. *The Musical Brain.* BrainFood Productions. 2013.

Saunders, Laura I. *Your Brain and Music: The Cognitive Effects of Music Education on the Brain.* Higher Purpose Publishing. 2017.

Acknowledgements

Most of all, I acknowledge all of those who have taught me science and music throughout my life. I also acknowledge my own students at Washington University, who encouraged me to write this book. Finally, I acknowledge and thank dramatist Mr. Larry Marsh, clarinetist Ms. Tina Ward, physicist Dr. Kenneth B. Freese, and neuroscientist Dr. Joseph C. Griffis for their critical readings of portions of this book, though any errors remain entirely my responsibility.

About the Author

Photo by Glenn Miller, Clayton Photos

Michael J. Montague earned his Ph.D. in cellular and molecular biology from The University of Michigan in 1974. Following post-doctoral studies at Stanford University, Dr. Montague began a 40-year research career with several major corporations, working on both the improvement of agricultural productivity and in pharmaceutical discovery, and additionally as a research manager, science communicator, and lecturer. During his career, he has enjoyed the immense privilege of membership on teams of research scientists who made major contributions that now benefit billions of people globally. At the same time, he worked as a card-carrying union musician, most often in the evenings, performing a wide variety of types of music, as well as arranging and composing. He has studied performance and music theory intensively with several excellent symphony musicians and teachers at Webster University in St. Louis and is most grateful for that valuable instruction. In addition, he has published original research on the acoustics of woodwind instruments.

Index

M

P

Q

R

S

T

W

Made in the USA
Las Vegas, NV
10 July 2026

50702124R10122